Hydrothermal Reduction of Carbon Dioxide to Low-Carbon Fuels

ELECTROCHEMICAL ENERGY STORAGE AND CONVERSION

Series Editor: Jiujun Zhang
National Research Council Institute for Fuel Cell Innovation
Vancouver, British Columbia, Canada

Published Titles

Hydrothermal Reduction of Carbon Dioxide to Low-Carbon Fuels
Fangming Jin

Carbon Nanomaterials for Electrochemical Energy Technologies: Fundamentals and Applications
Shuhui Sun, Xueliang Sun, Zhongwei Chen, Jinli Qiao, David P. Wilkinson, and Jiujun Zhang

Redox Flow Batteries: Fundamentals and Applications
Huamin Zhang, Xianfeng Li, and Jiujun Zhang

Electrochemical Energy: Advanced Materials and Technologies
Pei Kang Shen, Chao-Yang Wang, San Ping Jiang, Xueliang Sun, and Jiujun Zhang

Electrochemical Polymer Electrolyte Membranes
Jianhua Fang, Jinli Qiao, David P. Wilkinson, and Jiujun Zhang

Electrochemical Supercapacitors for Energy Storage and Delivery: Fundamentals and Applications
Aiping Yu, Victor Chabot, and Jiujun Zhang

Photochemical Water Splitting: Materials and Applications
Neelu Chouhan, Ru-Shi Liu, and Jiujun Zhang

Metal–Air and Metal–Sulfur Batteries: Fundamentals and Applications
Vladimir Neburchilov and Jiujun Zhang

Electrochemical Reduction of Carbon Dioxide: Fundamentals and Technologies
Jinli Qiao, Yuyu Liu, and Jiujun Zhang

Electrolytes for Electrochemical Supercapacitors
Cheng Zhong, Yida Deng, Wenbin Hu, Daoming Sun, Xiaopeng Han, Jinli Qiao, and Jiujun Zhang

Solar Energy Conversion and Storage: Photochemical Modes
Suresh C. Ameta and Rakshit Ameta

Lead-Acid Battery Technologies: Fundamentals, Materials, and Applications
Joey Jung, Lei Zhang, and Jiujun Zhang

Lithium-Ion Batteries: Fundamentals and Applications
Yuping Wu

Graphene: Energy Storage and Conversion Applications
Zhaoping Liu and Xufeng Zhou

Proton Exchange Membrane Fuel Cells
Zhigang Qi

Hydrothermal Reduction of Carbon Dioxide to Low-Carbon Fuels

Edited by
Fangming Jin

CRC Press
Taylor & Francis Group
Boca Raton London New York

CRC Press is an imprint of the
Taylor & Francis Group, an **informa** business

CRC Press
Taylor & Francis Group
6000 Broken Sound Parkway NW, Suite 300
Boca Raton, FL 33487-2742

First issued in paperback 2019

© 2018 by Taylor & Francis Group, LLC
CRC Press is an imprint of Taylor & Francis Group, an Informa business

No claim to original U.S. Government works

ISBN-13: 978-1-4987-3183-6 (hbk)
ISBN-13: 978-0-367-87371-4 (pbk)

This book contains information obtained from authentic and highly regarded sources. Reasonable efforts have been made to publish reliable data and information, but the author and publisher cannot assume responsibility for the validity of all materials or the consequences of their use. The authors and publishers have attempted to trace the copyright holders of all material reproduced in this publication and apologize to copyright holders if permission to publish in this form has not been obtained. If any copyright material has not been acknowledged please write and let us know so we may rectify in any future reprint.

Except as permitted under U.S. Copyright Law, no part of this book may be reprinted, reproduced, transmitted, or utilized in any form by any electronic, mechanical, or other means, now known or hereafter invented, including photocopying, microfilming, and recording, or in any information storage or retrieval system, without written permission from the publishers.

For permission to photocopy or use material electronically from this work, please access www.copyright.com (http://www.copyright.com/) or contact the Copyright Clearance Center, Inc. (CCC), 222 Rosewood Drive, Danvers, MA 01923, 978-750-8400. CCC is a not-for-profit organization that provides licenses and registration for a variety of users. For organizations that have been granted a photocopy license by the CCC, a separate system of payment has been arranged.

Trademark Notice: Product or corporate names may be trademarks or registered trademarks, and are used only for identification and explanation without intent to infringe.

Visit the Taylor & Francis Web site at
http://www.taylorandfrancis.com

and the CRC Press Web site at
http://www.crcpress.com

Contents

Editor vii
Contributors ix

Chapter 1 Water under High-Temperature and High-Pressure Conditions and Some Special Reactions under Hydrothermal Conditions 1

Zheng Shen, Wei Zhang, Xu Zeng, Fangming Jin, Guodong Yao, and Yuanqing Wang

Chapter 2 Catalytic Hydrothermal Reactions for Small Molecules Activation 23

Yuanqing Wang and Fangming Jin

Chapter 3 Hydrothermal Water Splitting for Hydrogen Production with Other Metals 37

Xu Zeng, Heng Zhong, Guodong Yao, and Fangming Jin

Chapter 4 Hydrothermal Water Splitting for Hydrogen Production with Iron ... 47

Xu Zeng, Heng Zhong, Guodong Yao, and Fangming Jin

Chapter 5 Hydrothermal CO_2 Reduction with Iron to Produce Formic Acid 61

Jia Duo, Guodong Yao, Fangming Jin, and Heng Zhong

Chapter 6 Hydrothermal Reduction of CO_2 to Low-Carbon Compounds 79

Ge Tian, Chao He, Ziwei Liu, and Shouhua Feng

Chapter 7 Hydrothermal CO_2 Reduction with Zinc to Produce Formic Acid 91

Yang Yang, Guodong Yao, Binbin Jin, Runtian He, Fangming Jin, and Heng Zhong

Chapter 8 Autocatalytic Hydrothermal CO_2 Reduction with Manganese to Produce Formic Acid 109

Lingyun Lyu, Fangming Jin, and Guodong Yao

Chapter 9 Autocatalytic Hydrothermal CO_2 Reduction with Aluminum to Produce Formic Acid 127

Binbin Jin, Guodong Yao, Fangming Jin, and Heng Zhong

Chapter 10 Cu-Catalyzed Hydrothermal CO_2 Reduction with Zinc to Produce Methanol 141

Zhibao Huo, Dezhang Ren, Guodong Yao, Fangming Jin, and Mingbo Hu

Chapter 11 Hydrothermal Reduction of CO_2 with Glycerine 153

Zheng Shen, Minyan Gu, Meng Xia, Wei Zhang, Yalei Zhang, and Fangming Jin

Chapter 12 Hydrothermal Reduction of CO_2 with Compounds Containing Nitrogen 185

Guodong Yao, Feiyan Chen, Jia Duo, Fangming Jin, and Heng Zhong

Chapter 13 Perspectives and Challenges of CO_2 Hydrothermal Reduction 199

Ligang Luo, Fangming Jin, and Heng Zhong

Index 205

Editor

Fangming Jin, PhD, is a distinguished professor in the School of Environmental Science and Engineering at Shanghai Jiao Tong University, China. She earned her PhD from Tohoku University, Japan. From 2007 to 2010, she was promoted to professor at Tongji University, China and also to chair professor sponsored by the Chang Jiang Scholar Program (administered by the Ministry of Education of the People's Republic of China). In 2010, she moved to Tohoku University as a professor, and then in July 2011, she moved to Shanghai Jiao Tong University and received an honor of "Recruitment Program of Global Experts" Talents in Shanghai. Her research focuses on the application of hydrothermal reactions for the conversion of CO_2 and biomass into fuels and chemicals, which aims to explore a potentially useful technology for the improvement of the carbon cycle by mimicking nature. She has authored more than 200 scientific publications, including peer-reviewed papers, patents, and book chapters, and gave about 30 plenary/keynote/invited presentations. Prof. Jin is also a visiting professor at Tohoku University and a Fellow of the Graduate School of Environmental Studies at Tohoku University.

Contributors

Feiyan Chen
School of Environmental Science and Engineering
State Key Laboratory of Metal Matrix Composites
Shanghai Jiao Tong University
Shanghai, China

Jia Duo
School of Environmental Science and Engineering
State Key Laboratory of Metal Matrix Composites
Shanghai Jiao Tong University
Shanghai, China

Shouhua Feng
State Key Laboratory of Inorganic Synthesis and Preparative Chemistry
College of Chemistry, Jilin University
Changchun, China

Minyan Gu
State Key Laboratory of Pollution Control and Resource Reuse
Tongji University
Shanghai, China

Chao He
Department of Earth and Planetary Sciences
Johns Hopkins University
Baltimore, Maryland

Runtian He
School of Environmental Science and Engineering
State Key Laboratory of Metal Matrix Composites
Shanghai Jiao Tong University
Shanghai, China

Mingbo Hu
State Key Laboratory of Pollution Control and Resources Reuse
College of Environmental Science and Engineering
Tongji University
Shanghai, China

Zhibao Huo
School of Environmental Science and Engineering
Shanghai Jiao Tong University
Shanghai, China

Binbin Jin
School of Environmental Science and Engineering
State Key Laboratory of Metal Matrix Composites
Shanghai Jiao Tong University
Shanghai, China

Fangming Jin
School of Environmental Science and Engineering
State Key Laboratory of Metal Matrix Composites
Shanghai Jiao Tong University
Shanghai, China

and

Graduate School of Environmental Studies
Tohoku University
Sendai, Japan

Ziwei Liu
MRC Laboratory of Molecular Biology
Cambridge Biomedical Campus
Cambridge, United Kingdom

Ligang Luo
School of Environmental Science and
 Engineering
State Key Laboratory of Metal Matrix
 Composites
Shanghai Jiao Tong University
Shanghai, China

Lingyun Lyu
School of Environmental Science and
 Engineering
State Key Laboratory of Metal Matrix
 Composites
Shanghai Jiao Tong University
Shanghai, China

Dezhang Ren
School of Environmental Science and
 Engineering
Shanghai Jiao Tong University
Shanghai, China

Zheng Shen
National Engineering Research Center
 for Facilities Agriculture
Institute of Modern Agricultural
 Science and Engineering
Tongji University
Shanghai, China

Ge Tian
State Key Laboratory of Inorganic
 Synthesis and Preparative Chemistry
College of Chemistry, Jilin University
Changchun, China

Yuanqing Wang
RIKEN Research Cluster for Innovation
 Nakamura Laboratory
Wako
Saitama, Japan

and

Tongji University
Shanghai, China

and

Fritz Haber Institute of the Max Planck
 Society
Berlin, Germany

Meng Xia
State Key Laboratory of Pollution
 Control and Resource Reuse
Tongji University
Shanghai, China

Yang Yang
School of Environmental Science and
 Engineering
State Key Laboratory of Metal Matrix
 Composites
Shanghai Jiao Tong University
Shanghai, China

Guodong Yao
School of Environmental Science and
 Engineering
State Key Laboratory of Metal Matrix
 Composites
Shanghai Jiao Tong University
Shanghai, China

Contributors

Xu Zeng
National Engineering Research Center for Facilities Agriculture
Institute of Modern Agricultural Science and Engineering
Tongji University
Shanghai, China

Wei Zhang
National Engineering Research Center for Facilities Agriculture
Institute of Modern Agricultural Science and Engineering
Tongji University
Shanghai, China

Yalei Zhang
State Key Laboratory of Pollution Control and Resource Reuse
Tongji University
Shanghai, China

Heng Zhong
Research Institute for Chemical Process Technology
National Institute of Advanced Industrial Science and Technology
Sendai, Japan

1 Water under High-Temperature and High-Pressure Conditions and Some Special Reactions under Hydrothermal Conditions

Zheng Shen, Wei Zhang, Xu Zeng, Fangming Jin, Guodong Yao, and Yuanqing Wang

CONTENTS

1.1 Introduction	1
1.2 Ion Product	2
1.3 Water Density	2
1.4 Dielectric Constant	3
1.4.1 Dielectric Constant of HTW	5
1.5 Hydrogen Bonding	6
1.6 Hydrolysis	11
1.7 Isomerization	12
1.8 Dehydration	13
1.9 Retro Aldol Reaction	14
1.10 Decarboxylation and Decarbonylation	15
References	16

1.1 INTRODUCTION

For water under high-temperature and high-pressure conditions, there are diverse terminologies that have been used in the literature. For example, high-temperature water (HTW) is defined as liquid water above 200°C [1]. Hot compressed water can also be used to denote water above 200°C and at sufficiently high pressure [2]. Water can be divided into subcritical water (below its critical point) and supercritical water (above its critical point) based on its critical point (T_c = 373°C, P_c = 22.1 MPa). The lower limit of temperature of subcritical water can be 100°C in the liquid state [3]. The terminology "near-critical water" is also often employed [4]. Aqueous phase processing is employed in the liquid water at temperatures of 200–260°C and at pressures of 10–50 bar to produce H_2, CO, and light alkanes from sugar-derived feedstocks [5]. The terminology "hydrothermal," which is originally from geology, has been more broadly and popularly used in literatures to refer to the reaction medium of high-temperature and high-pressure water. According to the different main products, it can be divided into hydrothermal carbonization (usually conducted at 100–200°C) [6], hydrothermal liquefaction (often at 200–350°C) [7], and hydrothermal gasification (often at 350–750°C) [8]. Therefore, in this chapter, the terminology "hydrothermal" will be mostly adopted to denote water above 100°C and 0.1 MPa, including subcritical and supercritical water.

The products distribution from hydrothermal biomass conversion, including gas, liquid, and solid, mainly depends on the properties of water at different states. Two competing reaction mechanisms are present: an ionic or polar reaction mechanism typical of liquid-phase chemistry at low temperature and a free radical reaction mechanism typical of gas-phase reactions at high temperature [9,10]. The latter radical reactions are preferred, leading to gas formation [11]. In addition, molecular reaction, which is different from ionic and radical reactions, is molecular rearrangement enhanced by coordination with water and proceeds around the critical region of water [12].

Herein, in the following sections, the representative properties of water under high-temperature and high-pressure conditions will be introduced, such as ion product, density, dielectric constant, and hydrogen bonding, and some special reactions under hydrothermal conditions will be discussed.

1.2 ION PRODUCT

The product of the concentrations of H^+ and OH^- in the water is denoted as the ion product (K_w), which is also called self-ionization constant (the unit of which is mol^2/kg^2). As the temperature increases, the ion product of water increases from $K_w = 10^{-14}$ mol^2/kg^{-2} at room temperature to approximately 10^{-11} mol^2/kg^{-2} at around 300°C at constant pressure (250 bar) [2]. Above the critical temperature, the ion product decreases sharply with increasing temperature [2]. In the ranges when water has a bigger K_w value, water may show enhanced acidic or basic catalytic activity for reactions owing to the high concentration of H^+ and OH^- ions [7]. Furthermore, it is expected to get a higher yield of target chemicals by adding minimal amounts of either acid or base catalysts. Antal et al. proposed that the ionic reactions are favored at $K_w > 10^{-14}$ and that free radical reactions are favored at $K_w < 10^{-14}$ [13]. In this section, five classes of reactions that often take place in the conversion of biomass are discussed, with one typical example to show the influence of ion product of water in the acid- or base-catalyzed reaction.

1.3 WATER DENSITY

Water density is another important property that can be varied greatly with temperature and pressure under hydrothermal conditions. Water density decreases with the increase in temperature at constant pressure. For example, water density decreases from about 800 kg/m³ (like liquid phase) to about 150 kg/m³ (like gas phase) without phase change as the temperature increases from 300°C to 450°C. Meanwhile, water density controlled by temperature and pressure can be related to ion product by Equation 1.1 using a fitting method proposed by Marshall and Franck [14].

$$\log K_w = A + \frac{B}{T} + \frac{C}{T^2} + \frac{D}{T^3} + \left(E + \frac{F}{T} + \frac{G}{T^2} \right) \log \rho, \quad (1.1)$$

where T is temperature in Kelvin, ρ is density in g/cm³, and A to G are fitting parameters. This result indicates that the chemistry of biomass conversion can also be controlled by water density. However, it is not to say that water density affects the reaction mechanism only by changing the ion product of water. Water density changes can reflect the changes of water at the molecular level, such as solvation effect, hydrogen bonding, polarity, dielectric strength, molecular diffusivity, and viscosity, which will influence the chemistry inside [15]. In supercritical water, the reaction mechanism varies from a reaction atmosphere that favors radical reaction to one that favors ionic reactions dictated by the water density [16]. Experimental data showed that reactions seemed to proceed via ionic pathways in the high-density water while radical reactions seemed to be the main reaction pathways in the less dense supercritical water [9]. Westacott et al. [17,18] investigated *tert*-butyl chloride dissociation in supercritical water by computational methods and found that water density affected the competition between ionic and radical reaction mechanisms. The ionic heterolytic dissociation was preferred over the radical homolytic dissociation when water density was larger than 0.03 g/cm³ [17,18]. In this section, different reaction mechanisms via ionic or radical pathways affected by water density were introduced using different feedstocks.

1.4 DIELECTRIC CONSTANT

The ratio of the permittivity of a substance to the permittivity of free space is denoted as the dielectric constant. The dielectric constant of water under ambient condition is 78.5. Water under this condition could be used as good solvent for the polar materials. However, it cannot be used to dissolve hydrocarbon and gases. The dielectric constant of water as a function of temperature can be seen in Figure 1.1 [19]. As shown in Figure 1.1, the dielectric constant of water decreases sharply with the increase of water temperature. HTW under subcritical and supercritical conditions behaves like many organic solvents that can dissolve organic compounds completely forming a single fluid phase. The advantages of a single supercritical phase reaction medium are that higher concentrations of reactants can often be attained and no interphase mass transport processes hindering the reaction rates were indispensable.

As a consequence of the lack of data, attempts to estimate the properties of aqueous species at high temperature and/or high pressure rely on the estimated or extrapolated dielectric constant values [20]. The dielectric constant dependence on the pressure, proposed by Bradley and Pitzer [21], can be seen in Figure 1.2. Bradley used an equation suggested by Tait in 1880 for volumetric data. As shown in this

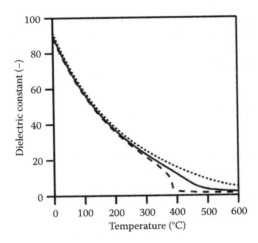

FIGURE 1.1 Dielectric constant of water as a function of temperature. Dashed line: 25 MPa; solid line: 50 MPa; dotted line: 100 MPa. Reprinted from *Journal of Bioscience and Bioengineering*, 117, Akizuki M, Fujii T, Hayashi R, Oshima Y, Effects of water on reactions for waste treatment, organic synthesis, and bio-refinery in sub- and supercritical water, 10–18, Copyright 2014, with permission from Elsevier.

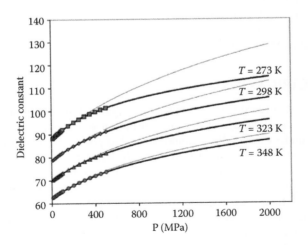

FIGURE 1.2 Dielectric constant of water as a function of pressure at constant temperatures (273 K, 298 K, 323 K, and 348 K) (fine lines: Bradley's equation [21]; thick lines: adjusted values extracted from the *International Association for the Properties of Water and Steam* [22]). Reprinted with permission from Bradley DJ, Pitzer KS (1979) Thermodynamics of electrolytes. 12. Dielectric properties of water and Debye-Hueckel parameters to 350°C and 1 kbar. *Journal of Physical Chemistry* 83 (12):1599–1603. Copyright 2013 American Chemical Society.

figure, at constant temperature, the dielectric constant values increased linearly with the increase of pressure. It should be noted that the original Bradley equation does not reproduce adequately the data available from the *International Association for the Properties of Water and Steam* [22] used in Figure 1.2 for pressure above 400 MPa, particularly at and above 323 K. However, the trends are similar, which can approximately represent the change of dielectric constant with different pressure.

1.4.1 Dielectric Constant of HTW

The dielectric constant of HTW attracted much attention. Islam et al. compared the dielectric constant (ε) of superheated water at different temperature and pressure, as shown in Table 1.1 [23]. The dielectric constant values of water decreased with the increase of temperature from 44 at 150°C to 2 at 350°C. These values are between those of organic solvent ethanol ($\varepsilon = 24$ at 25°C) and methanol ($\varepsilon = 33$ at 25°C), indicating that superheated water can be used as an organic solvent. Moreover, superheated water is readily available, nontoxic, reusable, and very low in cost as well as environmentally friendly. Therefore, superheated water can be used as an alternative cleaning technology, instead of using organic solvents or toxic and strong aqueous liquid media, for example, the extraction of dioxins [24], pesticides [25], polychlorinated biphenyls [26], and polycyclic aromatic hydrocarbons (PAHs) [27]. Lagadec et al. reported that the optimum subcritical water extraction was at 275°C in 35 min for all low- and high-molecular-weight PAHs from contaminated manufactured gas plant soil [25]. Moreover, it can also be used to determine a superior instant analytical technique (using gas chromatography oven as heater) by using organic solvent [27]. However, a complete extraction technology with shorter extraction time at a temperature range (from 100°C to 300°C) using subcritical water for industrial application has not been determined; therefore, an additional study is necessary [23].

Notably, the dielectric constant of supercritical water is very special, because the dielectric constant under this condition is much lower, and the number of hydrogen bonds is much smaller and their strength is much weaker. Supercritical water above 374°C and 221 bar shows that water is greatly diminished—frequently less than reduced local molecular ordering and less effective hydrogen bonding as characterized by its lower dielectric constant (about 1 to 3) [28]. As a result, supercritical water

TABLE 1.1
Dielectric Constant (ε) of Subcritical Water and Common Organic Solvent

ε (at Subcritical Water °C)	ε of Common Organic Solvent at 25°C
44 (150)	1.9 (*n*-hexane)
35 (200)	21 (acetone)
27 (250)	24 (ethanol)
20 (300)	33 (methanol)
2 (350)	39 (acetonitrile)

TABLE 1.2
Dielectric Constant and Density of Water at Some Supercritical Conditions

Temperature (°C)	Pressure (MPa)	Density (g/cm³)	Dielectric Constant
400	25	0.17	2.4
400	30	0.35	5.9
500	25	0.09	1.5
500	30	0.12	1.7
350	25	0.63	14.85

behaves like many organic solvents so that organic compounds have complete miscibility with supercritical water. Moreover, gases are also miscible in supercritical water; thus, a supercritical water reaction environment provides an opportunity to conduct the chemical reactions in a single fluid phase that would otherwise occur in a multiphase system under conventional conditions [29]. Therefore, supercritical water exhibits considerable characters of solvent, which can dissolve nonpolar materials and gas, and the characters of easy diffusion and motion [30]. The dielectric constant of supercritical water corresponds to the value of polar solvent under ambient conditions. The dielectric constant of ambient water varies continuously over a much larger range in the supercritical state. This variation offers the possibility of using pressure and temperature to influence the properties of the reaction medium. Therefore, it is possible for the formation of a C–C bond with organometallic catalytic reactions, which always needs organic solvent. Gomez-Briceno et al. compared the dielectric constant of water at different supercritical conditions, 400°C and 500°C, and two pressures values, 25 and 30 MPa, as shown in Table 1.2 [31]. The data showed that the dielectric constant decreased significantly with the decrease of temperature. However, the influence was very small.

Water with a large dielectric constant will exhibit a strong effect with the microstructure of water and eventually influence the reaction [2]. The large dielectric constant indicates that substances whose molecules contain ionic bonds tend to dissociate in water, yielding solutions containing ions. This occurs because water as a solvent opposes the electrostatic attraction between positive and negative ions that would prevent ionic substances from dissolving. These separated ions become surrounded by the oppositely charged ends of the water dipoles and become hydrated. This ordering tends to be counteracted by the random thermal motions of the molecules. Water molecules are always associated with each other through as many as four hydrogen bonds, and this ordering of the structure of water greatly resists the random thermal motions. Indeed, it is this hydrogen bonding that is responsible for its large dielectric constant.

1.5 HYDROGEN BONDING

Because of hydrogen bonding, WHTP (water under high temperature and pressure conditions) exhibits many unique properties that are quite different from those of ambient liquid water.

The number of hydrogen bonds per water molecule at different temperature and density is shown in Figure 1.3. It can be concluded from Figure 1.3 that with increasing temperature and decreasing density, hydrogen bonding in water becomes weaker and less persistent [32]. For example, water at 673 K and ~0.5 g/cm^3 retains 30–45% of the hydrogen bonding that exists at ambient conditions, whereas water at 773 K and ~0.1 g/cm^3 retains 10–14% [33]. The hydrogen bonding network in ambient liquid water exists in the form of infinite percolating large clusters of hydrogen-bonded water molecules, but the hydrogen bonding network in WHTP exists in the form of small clusters of hydrogen-bonded water molecules [32,34–36]. In general, the average cluster size of hydrogen-bonded water molecules decreases with increasing temperature and decreasing density. For instance, the size of most of the clusters at 773–1073 K and 0.12–0.66 g/cm^3 consists of five water molecules or less, although there exist a small number of clusters that are as large as about 20 water molecules [32,35,36]. These results show that less hydrogen bonding results in much less order in WHTP than ambient liquid water and that individual water molecules can participate in elementary reaction steps as a hydrogen source or catalyst during hydrothermal conversion of biomass into high-valued chemicals.

Many studies have reported on water molecules supplying hydrogen atoms that participate in reactions such as the steam reforming of glucose [37,38] and biomass [39,40]; the pyrolysis of alkyldiammonium dinitrate [41]; the oxidation of methylene chloride [42], lactic acid [43], and carbon monoxide [44–46]; the hydrogenation of dibenzothiophene [47] and heavy oils [48]; the co-liquefaction of coal and cellulose [49]; and the alcohol-mediated reduction of CO_2 and $NaHCO_3$ into formate [50,51]. They produced hydrogen in situ by partially oxidizing the organic compounds to generate carbon monoxide, which then underwent the water–gas shift reaction ($CO + H_2O \leftrightarrow CO_2 + H_2$). The authors proposed that the reactive intermediate generated by the water–gas shift reaction

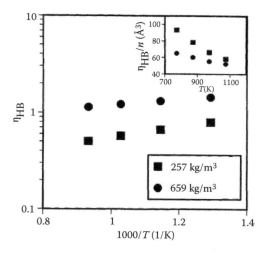

FIGURE 1.3 Number of hydrogen bonds per water molecule. [From Mizan TI, Savage PE, Ziff RM (1996) Temperature dependence of hydrogen bonding in supercritical water. *Journal of Physical Chemistry* 100 (1):403–408.]

was the actual hydrogenation agent and not the hydrogen molecule itself. As shown in Figure 1.4, our recent study found that CO_2 or $NaHCO_3$ could be transformed into formate by alcohol-mediated reduction under hydrothermal alkaline conditions [50,51].

Hydrogen–deuterium exchange data also provide evidence for hydrogen supply by water as deuterium was found to be incorporated into the products of hydrocarbon pyrolyses in supercritical D_2O [52,53]. More recently, in order to discover the reaction mechanism for the production of hydrogen and lactic acid from glycerol under alkaline hydrothermal conditions, we identified the different intermediates involved during reactions by investigating the water solvent isotope effect with ^{1}H-NMR, ^{2}H-NMR, LC–MS, and Gas-MS analyses as shown in Figure 1.5 [53]. The results from solvent isotope studies showed that (1) almost all of the H atoms on the β-C

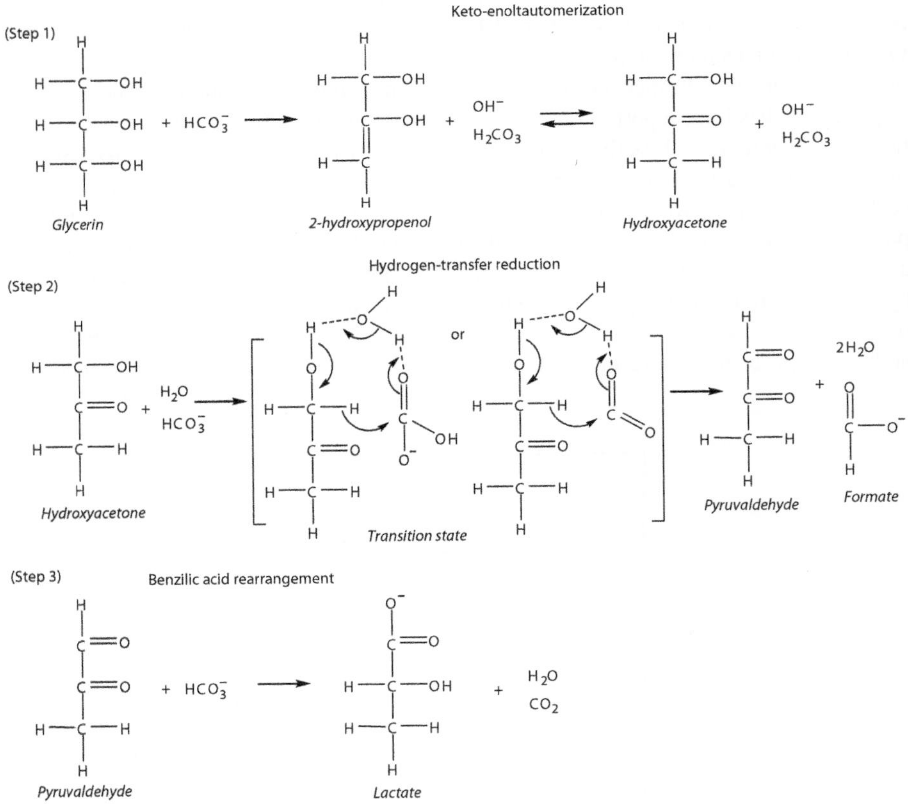

FIGURE 1.4 The proposed pathway of the hydrogen-transfer reduction of $NaHCO_3$ with glycerine. [From Shen Z, Zhang YL, Jin FM (2012) The alcohol-mediated reduction of CO_2 and $NaHCO_3$ into formate: a hydrogen transfer reduction of $NaHCO_3$ with glycerine under alkaline hydrothermal conditions. *Rsc Adv* 2 (3):797–801.]

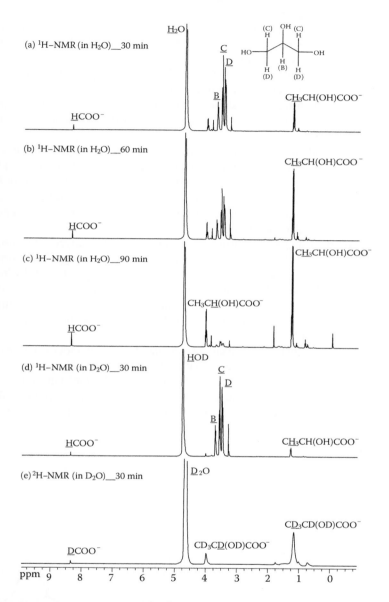

FIGURE 1.5 ^{1}H–NMR spectra for the solution after the hydrothermal reaction of 0.33 M glycerol at 300°C with 1.25 M NaOH in H_2O for (a) 30 min, (b) 60 min, (c) 90 min, and (d) ^{1}H-NMR and (e) ^{2}H-NMR spectra with 1.25 M NaOD in D_2O for 30 min. [From Zhang YL, Shen Z, Zhou XF, Zhang M, Jin FM (2012) Solvent isotope effect and mechanism for the production of hydrogen and lactic acid from glycerol under hydrothermal alkaline conditions. *Green Chemistry* 14 (12):3285–3288.]

of lactic acid were exchanged by D_2O, suggesting that the hydroxyl (–OH) group on the 2-C of glycerol was first transformed into a carbonyl (C=O) group and then converted back into an –OH group to form lactic acid; (2) a large amount of D was found in the produced hydrogen gas, which shows that the water molecules acted as a reactant; and (3) the percentage of D in the produced hydrogen gas was far more than 50%, which straightforwardly showed that acetol was formed in the first place as the most probable intermediate by undergoing a dehydration reaction rather than a dehydrogenation reaction.

The natural abundance of hydronium and hydroxide ions suggests that some acid- and base-catalyzed reaction may proceed in HTW in the absence of an added catalyst [54–67]. Alcohol dehydration is nominally catalyzed by either acid or base in the presence of added catalysts. In WHTP, however, experimental data suggest that the dominant mechanism is acid catalysis and the dehydration reactivity depends on the structure of the alcohol [54–60].

Experimental data suggest that water molecules can also catalyze a reaction by directly participating in the transition state and reducing its energy. This form of catalysis is important for reactions involving some types of intramolecular hydrogen transfer. For example, Klein, Brill, and coworkers proposed a type of water catalysis for the intramolecular hydrogen-transfer step during the conversion of nitroaniline to benzofurazan as shown in Figure 1.6 [64] and the decarboxylation of acetic acid derivatives in WHTP [65].

Arita et al. reported that hydrogen can be generated by an ethanol oxidation reaction catalyzed by water molecules and that half of the produced hydrogen could come from the water in accordance with the proposed reaction mechanism in Figure 1.7 [61]. Moreover, Takahashi et al. suggested that water molecules played significant catalytic roles in ethanol oxidation reactions based on ab initio density functional theory calculations [66].

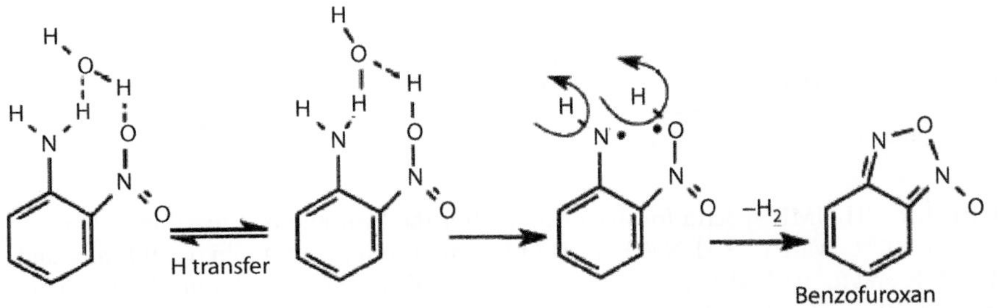

FIGURE 1.6 Water catalysis for the intramolecular hydrogen transfer during the conversion of nitroaniline to benzofurazan. [From Wang XG, Gron LU, Klein MT, Brill TB (1995) The influence of high-temperature water on the reaction pathways of nitroanilines. *J Supercrit Fluids* 8 (3):236–249.]

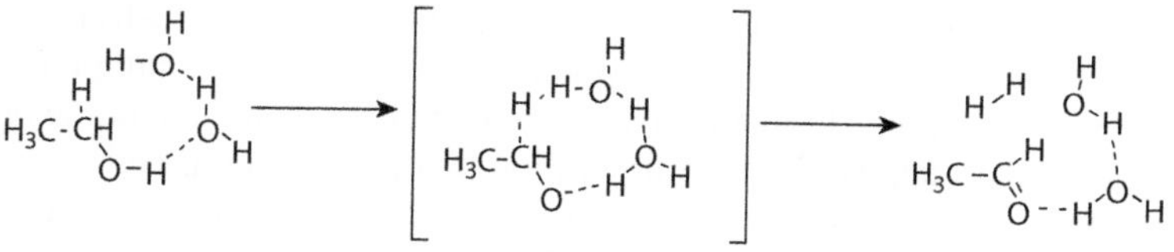

FIGURE 1.7 Proposed transition state consisting of an ethanol molecule and two water molecules in supercritical water without catalyst. [From Arita T, Nakahara K, Nagami K, Kajimoto O (2003) Hydrogen generation from ethanol in supercritical water without catalyst. *Tetrahedron Letters* 44 (5):1083–1086.]

1.6 HYDROLYSIS

As shown in Figure 1.8, hydrolysis is one of the major (and usually initial) reactions that occur during conversion of biomass in which glycosidic bonds between sugar units are cleaved to form simple sugars such as glucose and partially hydrolyzed oligomers. Hydrolysis can happen both in acid- and base-catalyzed reactions, while the former reaction condition (acidic) is more often adopted because base catalysis leads to more side reactions [68,69]. The hydrolysis of cellulose to glucose is a widely investigated reaction in biomass conversion because cellulose is the major component of plant biomass and the product glucose is a very important intermediate [70]. Under hydrothermal conditions, cellulose reacts with water and is hydrolyzed into glucose or other monomers proceeding through C–O–C bond cleavage and accompanied by further degradation. Three possible reaction paths of cellobiose hydrolysis are demonstrated, including acid-, base-, and water-catalyzed ways [69]. Acid hydrolysis proceeds through the formation of a conjugated acid followed by the glycosidic bond cleavage and leads to the two glucose units. In the base pathway, the OH⁻ attacks at the anomeric carbon atom, renders the cleavage of the O bridge, and again yields the two glucose units. The water-catalyzed reaction is characterized by H_2O adsorption. Then, water and the glycosidic bond split simultaneously and form two glucoses again. Sasaki et al. [71,72] conducted cellulose decomposition experiments with a flow-type reactor from 290°C to 400°C at 25 MPa. Higher hydrolysis product yields (around 75%) were obtained in supercritical water than in subcritical water. The reason was attributed to the difference of reaction rate in the formation and degradation of oligomer or glucose. At a low-temperature region,

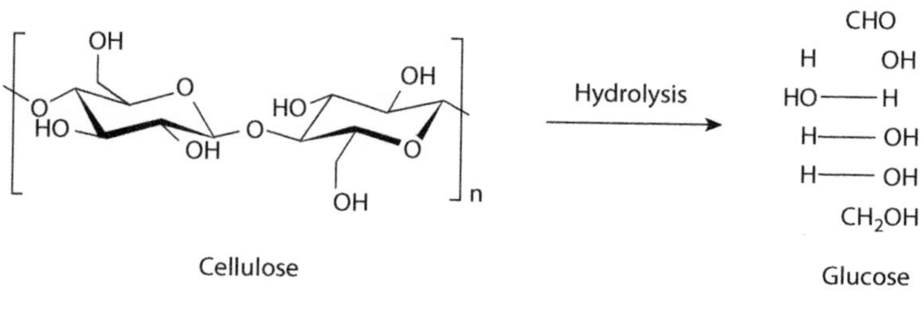

FIGURE 1.8 Hydrolysis of cellulose.

the glucose or oligomer conversion rate was much faster than the hydrolysis rate of cellulose. However, around the critical point, the hydrolysis rate jumped to more than an order of magnitude higher level and became faster than the glucose or oligomer decomposition rate. The direct observation by diamond anvil cell showed that the cellulose disappeared with a more than two orders of magnitude faster rate at 300–320°C than that estimated [72]. This phenomenon indicated that the presence of a homogeneous hydrolysis atmosphere caused by the dissolution of cellulose or hydrolyzed oligomers around the critical temperature thus resulted in the high cellulose hydrolysis rate. The additional acid catalysts including homogeneous and heterogeneous catalysts would also enhance the yield of glucose, which was around 50–80% [70]. The base catalyst might cause more side reactions [69] but could inhibit the formation of char, which was very crucial in the continuous flow reactor to prevent plug [73].

1.7 ISOMERIZATION

As shown in Figure 1.9, the isomerization between glucose and fructose is very common and has been considered as one key step in biomass conversion. The difference in their reactivity and selectivity for target materials makes the tunable transformation to a specific one (usually from glucose to fructose) highly desirable [74]. This reaction is typically catalyzed by the base catalyst, named Lobry de Bruyn–Alberda van Ekenstein transformation. The mechanism proceeds by deprotonation of alpha carbonyl carbon of glucose by base, resulting in the formation of a series of enolate intermediates. The overall process involves hydrogen transfer from C-2 to C-1 and from O-2 to O-1 of an alpha hydroxy aldehyde to form the corresponding ketone. Kabyemela et al. [75] found that the isomerization from fructose to glucose is negligible compared with its reversion under hydrothermal conditions because glucose and fructose have the same product distribution except for 1,6-anhydroglucose, which is not observed in the decomposition of fructose. Recently, Wang et al. [76] reported another Lewis acid-catalyzed pathway of isomerization via intramolecular hydride transfer for glucose–fructose. In addition to glucose–fructose isomerization, there is another important isomerization between glyceraldehydes and dihydroxyacetone under hydrothermal conditions [77].

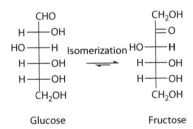

FIGURE 1.9 Isomerization between glucose and fructose.

1.8 DEHYDRATION

Dehydration reactions of biomass comprise an important class of reactions in the area of sugar chemistry. As shown in Figure 1.10, fructose can be dehydrated into hydroxymethylfurfural (HMF) with loss of three water molecules by acid-catalyzed reaction. Antal et al. [78] proposed that HMF is produced from fructose via cyclic intermediates. Recent studies confirmed that the HMF formation was from the acid-catalyzed dehydration of C6 sugars in the furanose form [79,80]. Hence, fructose, which contains 21.5% of furanose tautomers in aqueous solution, can be converted to HMF easier than glucose, which contains only 1% of furanose tautomers in aqueous solutions. The rehydration of HMF with two molecules of water would produce levulinic acid and formic acid [81]. Levulinic acid can be further converted into γ-valerolactone via hydrogenation with hydrogen [82], which can be converted into liquid alkenes in the molecular weight range appropriate for transportation fuel [83].

Asghari and Yoshida obtained the best yield of HMF (65%) from fructose achieved at a temperature of 513 K for a residence time of 120 s [84]. Since glucose is more common than fructose in biomass conversion, researchers usually adopt a two-step strategy to produce HMF from glucose: (1) isomerization of glucose into fructose catalyzed by base and (2) dehydration of fructose into HMF by acid [85]. Since water under high temperatures and pressures can play the role of both acid and base catalysts, a high yield of HMF can be obtained under hydrothermal conditions in one step. Takeuchi et al. [86] reported that the total highest yields of HMF and levulinic acid from glucose were about 50%, which occurred at 523 K for 5 min with H_3PO_4 as a catalyst, and the highest yield of levulinic acid was about 55% at 523 K for 5 min with HCl as a catalyst. For the three mineral acids (HCl, H_2SO_4, and H_3PO_4), it was found that both the pH and the nature of the acids had great influence on the decomposition pathway [84]. The order for the production of HMF using the three acids was as follows: $H_3PO_4 > H_2SO_4 > HCl$ [86]. On the contrary, the order for production of levulinic acid was as follows: $HCl > H_2SO_4 > H_3PO_4$ [86].

There are some drawbacks in the acid-catalyzed formation of HMF from fructose or glucose. Kinetic studies [87–89] showed that humins formation from glucose and HMF cannot be neglected. The activation energy of its formation from glucose and HMF was estimated at 51 and 142 kJ/mol, respectively, while dehydration of

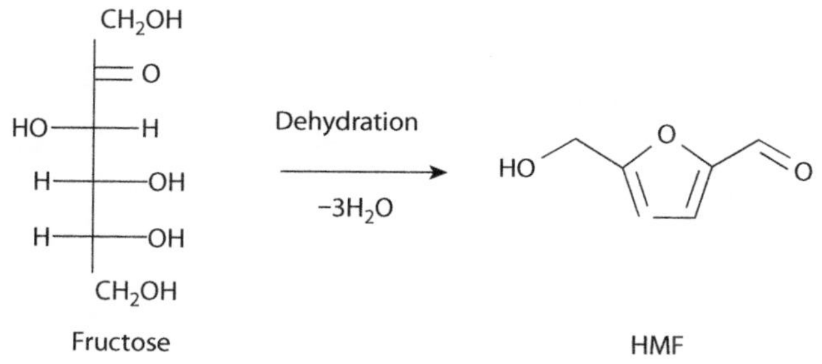

FIGURE 1.10 Dehydration of fructose into HMF.

glucose to HMF and rehydration of HMF to levulinic acid were 160 and 95 kJ/mol, respectively [89]. To minimize the formation of humins and enhance the selectivity toward HMF, a biphasic solution with water and organic phase was adopted that would continuously extract HMF as it is produced [90–93]. Yang et al. reported a 61% yield of HMF from glucose using a biphasic reactor of water/tetrahydrofuran with $AlCl_3 \cdot 6H_2O$ catalyst at 160°C [91].

1.9 RETRO ALDOL REACTION

Many researchers [67,71,75,77,94] have examined intermediate products for the hydrothermal degradation of glucose and cellulose at a reaction temperature of near 300°C. As shown in Figure 1.11, through these studies, it was revealed that some compounds containing three carbon atoms, such as glyceraldehyde, dihydroxyacetone, and pyruvaldehyde, were formed by the base catalytic role of HTW. Furthermore, there was isomerization occurring between glyceraldehyde and dihydroxyacetone followed by their subsequent dehydration to pyruvaldehyde [77]. The ketone (fructose) can undergo reverse aldol reaction by C3–C4 bond cleavage to form glyceraldehydes. These C3 carbon compounds were considered as the precursors of lactic acid from transformation of pyruvaldehyde [67]. On the other hand, the intermediates glycoaldehyde and erythrose were transformed from glucose by retro aldol reaction [71,75]. In organic chemistry, retro aldol reaction can usually be catalyzed by either an acid or a base. Experimental data suggested, however, that retro aldol reaction under hydrothermal conditions was base catalyzed [2]. Sasaki et al. [95] reported that the retro aldol reaction selectively proceeded at higher temperatures (above 673 K) and lower pressure (below 25 MPa). At a low temperature, the retro aldol reaction was preferred in an alkali environment [96].

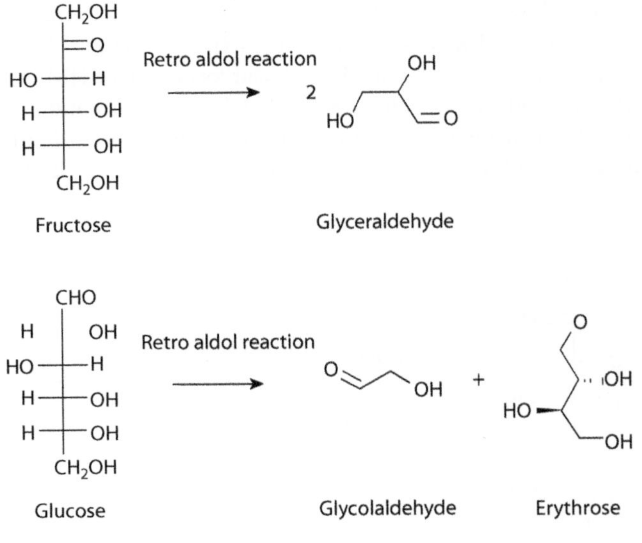

FIGURE 1.11 Retro aldol reaction of fructose and glucose.

These formed intermediates from C2-C3 or C3-C4 bond cleavage by reverse aldol reaction from hexoses can then be fast transformed into mainly lactic acid and other low molecular acid in which glyceraldehyde can produce a higher yield of lactic acid [94]. Lactic acid is a key chemical and acts as a building block for biodegradable lactic acid polymers with limited environmental impact. Yan et al. [97,98] showed that the addition of base catalyst [NaOH and $Ca(OH)_2$] can increase the yield of lactic acid. The highest yield of lactic acid from glucose was 27% with 2.5 M NaOH and 20% with 0.32 M $Ca(OH)_2$ at 300°C for 60 s [97]. A very recent study by Sánchez et al. [99] also found that the highest-yield lactic acid of 45% from corn cobs was obtained using 0.7 M $Ca(OH)_2$ at 300°C for 30 min. The reason that the base catalyst increased the yield of lactic acid can be attributed to the enhancement of the reaction pathway for lactic acid production discussed above. Another possible reason is that the lactate formed actually by alkaline solution prevents it from decomposition [100]. Compared with NaOH at lower alkaline concentration [95], $Ca(OH)_2$ more effectively promoted the production of lactic acid than NaOH at the same OH^- concentration. It is probably because Ca^{2+} was more capable than Na^+ in forming complexes with two oxygen atoms in the hexoses. When the concentration of $Ca(OH)_2$ increased from 0.32 to 0.4 M, it did not lead to an increase in lactic acid yield, while the optimum OH^- concentration for NaOH was 2.5 M. This difference can be attributed to the fact that the saturated solubility of NaOH is higher than that of $Ca(OH)_2$.

1.10 DECARBOXYLATION AND DECARBONYLATION [100–104]

$$HCOOH \rightarrow H_2 + CO_2 \text{ (decarboxylation)} \quad (1.2)$$

$$HCOOH \rightarrow H_2O + CO \text{ (decarbonylation)} \quad (1.3)$$

The reactions of formic acid play a key role in the chemistry of hydrothermal reaction, partly because it was the simplest acid and a product of many acid/base-catalyzed or oxidation reactions, and partly because itself or formate is considered to be the intermediate of water–gas shift reaction and reduction of carbon dioxide [105]. The understanding of its reactivity especially coupled with the properties of water will facilitate the researches on energy production and environment protection. As shown in Equations 1.2 and 1.3, for the decomposition of formic acid, there existed two competitive pathways: decarboxylation and decarbonylation. Early experimental results showed that, in the gas phase, decarbonylation dominated, but in the liquid phase, decarboxylation dominated [101,106]. Yu and Savage [101] conducted the formic acid decomposition experiments from 320°C to 500°C, at pressures from 18.3 to 30.4 MPa, and at 1.4- to 80-s reaction times. Conversion rates ranging from 38% to 100% were obtained with the major products of CO_2 and H_2. In their experiments, the decarbonylation product CO was also detected and the yields were always at least an order of magnitude lower than the yields of decarboxylation. The reason why decarboxylation dominated in the liquid phase can be explained by the presence of water as a homogeneous catalyst that can catalyze decarboxylation more than decarbonylation by a theoretical calculation [102]. The kinetic data also supported the assumption of a homogeneous reaction based on the consistency with the reaction rate law that was first order in formic acid [101]. However, Wakai et al. [103] indicated that the reactor wall might show a catalytic role in the formic acid decomposition, which is a heterogeneous reaction according to an NMR investigation. Compared with its acidic environment, Jin et al. [100] found that the addition of alkali could prevent the formic acid decomposition even with the presence of oxidant H_2O_2 at 250°C for 60 s.

REFERENCES

1. Akiya N, Savage PE (2002) Roles of water for chemical reactions in high-temperature water. *Chemical Reviews* 102 (8):2725–2750.
2. Kruse A, Dinjus E (2007) Hot compressed water as reaction medium and reactant—Properties and synthesis reactions. *The Journal of Supercritical Fluids* 39 (3):362–380.
3. Pourali O, Asghari FS, Yoshida H (2009) Sub-critical water treatment of rice bran to produce valuable materials. *Food Chemistry* 115 (1):1–7.
4. Watanabe M, Sato T, Inomata H, Smith RL, Arai K, Kruse A, Dinjus E (2004) Chemical reactions of C-1 compounds in near-critical and supercritical water. *Chemical Reviews* 104 (12):5803–5821.
5. Huber GW, Iborra S, Corma A (2006) Synthesis of transportation fuels from biomass: Chemistry, catalysts, and engineering. *Chemical Reviews* 106 (9):4044–4098.
6. Titirici MM, White RJ, Falco C, Sevilla M (2012) Black perspectives for a green future: Hydrothermal carbons for environment protection and energy storage. *Energy & Environ Sci* 5 (5):6796–6822.
7. Jin FM, Enomoto H (2011) Rapid and highly selective conversion of biomass into value-added products in hydrothermal conditions: Chemistry of acid/base-catalysed and oxidation reactions. *Energy & Environ Sci* 4 (2):382–397.
8. Kruse A, Bernolle P, Dahmen N, Dinjus E, Maniam P (2010) Hydrothermal gasification of biomass: Consecutive reactions to long-living intermediates. *Energy & Environ Sci* 3 (1):136–143.
9. Buhler W, Dinjus E, Ederer HJ, Kruse A, Mas C (2002) Ionic reactions and pyrolysis of glycerol as competing reaction pathways in near- and supercritical water. *The Journal of Supercritical Fluids* 22 (1):37–53.
10. Kruse A, Dinjus E (2007) Hot compressed water as reaction medium and reactant—2. Degradation reactions. *The Journal of Supercritical Fluids* 41 (3):361–379.
11. Kruse A (2008) Supercritical water gasification. *Biofuels Bioproducts & Biorefining* 2 (5):415–437.
12. Sato T, Sekiguchi G, Adschiri T, Arai K (2002) Ortho-selective alkylation of phenol with 2-propanol without catalyst in supercritical water. *Industrial & Engineering Chemistry Research* 41 (13):3064–3070.
13. Antal MJ, Brittain A, Dealmeida C, Ramayya S, Roy JC (1987) Heterolysis and homolysis in superitical water. *ACS Symposium Series* 329:77–86.
14. Marshall W, Franck E (1981) Ion product of water substance, 0–1000°C, 1–10,000 bars—New international formulation and its background. *Journal of Physical and Chemical Reference Data* 10 (2):295–304.

15. Peterson AA, Vogel F, Lachance RP, Froling M, Antal MJ, Tester JW (2008) Thermochemical biofuel production in hydrothermal media: A review of sub- and supercritical water technologies. *Energy and Environmental Science* 1 (1):32–65.
16. Katritzky AR, Nichols DA, Siskin M, Murugan R, Balasubramanian M (2001) Reactions in high-temperature aqueous media. *Chemical Reviews* 101 (4):837–892.
17. Westacott RE, Johnston KP, Rossky PJ (2001) Stability of ionic and radical molecular dissociation pathways for reaction in supercritical water. *Journal of Physical Chemistry B* 105 (28):6611–6619.
18. Westacott RE, Johnston KP, Rossky PJ (2001) Simulation of an SN1 reaction in supercritical water. *Journal of the American Chemical Society* 123 (5):1006–1007.
19. Akizuki M, Fujii T, Hayashi R, Oshima Y (2014) Effects of water on reactions for waste treatment, organic synthesis, and bio-refinery in sub- and supercritical water. *Journal of Bioscience and Bioengineering* 117:10–18.
20. Floriano WB, Nascimento MAC (2004) Dielectric constant and density of water as a function of pressure at constant temperature. *Brazilian Journal of Physics* 34:38–41.
21. Bradley DJ, Pitzer KS (1979) Thermodynamics of electrolytes. 12. Dielectric properties of water and Debye-Hueckel parameters to 350°C and 1 kbar. *Journal of Physical Chemistry* 83 (12):1599–1603.
22. Lide DR (1993) *CRC Handbook of Chemistry and Physics*. 74th edn. CRC Press, Boca Raton, FL.
23. Islam MN, Jo YT, Park JH (2012) Remediation of PAHs contaminated soil by extraction using subcritical water. *Journal of Industrial and Engineering Chemistry* 18 (5):1689–1693.
24. Hashimoto S, Watanabe K, Nose K, Morita M (2004) Remediation of soil contaminated with dioxins by subcritical water extraction. *Chemosphere* 54 (1):89–96.
25. Lagadec AJM, Miller DJ, Lilke AV, Hawthorne SB (2000) Pilot-scale subcritical water remediation of polycyclic aromatic hydrocarbon- and pesticide-contaminated soil. *Environmental Science & Technology* 34 (8):1542–1548.
26. Yang Y, Bowadt S, Hawthorne SB, Miller DJ (1995) Subcritical water extraction of polychlorinated-biphenyls from soil and sediment. *Analytical Chemistry* 67 (24):4571–4576.
27. Hawthorne SB, Yang Y, Miller DJ (1994) Extraction of organic pollutants from environmental solids with sub- and supercritical water. *Analytical Chemistry* 66 (18):2912–2920.
28. Smith KA, Griffith P, Harris JG, Herzog HJ, Howard JB, Latanision R, Peters WA (1995) Supercritical water oxidation: Principles and prospects. Paper presented at the Proceedings of the International Water Conference, Pittsburgh.
29. Savage PE (1999) Organic chemical reactions in supercritical water. *Chemical Reviews* 99 (2):603–621.
30. Eckert CA, Knutson BL, Debenedetti PG (1996) Supercritical fluids as solvents for chemical and materials processing. *Nature* 383 (6598):313–318.
31. Gomez-Briceno D, Blazquez F, Saez-Maderuelo A (2013) Oxidation of austenitic and ferritic/martensitic alloys in supercritical water. *The Journal of Supercritical Fluids* 78:103–113.
32. Mizan TI, Savage PE, Ziff RM (1996) Temperature dependence of hydrogen bonding in supercritical water. *Journal of Physical Chemistry* 100 (1):403–408.
33. Hoffmann MM, Conradi MS (1997) Are there hydrogen bonds in supercritical water? *Journal of the American Chemical Society* 119 (16):3811–3817.
34. Jedlovszky P, Brodholt JP, Bruni F, Ricci MA, Soper AK, Vallauri R (1998) Analysis of the hydrogen-bonded structure of water from ambient to supercritical conditions. *Journal of Chemical Physics* 108 (20):8528–8540.

35. Kalinichev AG, Churakov SV (1999) Size and topology of molecular clusters in supercritical water: A molecular dynamics simulation. *Chemical Physics Letters* 302 (5–6):411–417.
36. Mountain RD (1999) Voids and clusters in expanded water. *Journal of Chemical Physics* 110 (4):2109–2115.
37. Holgate HR, Meyer JC, Tester JW (1995) Glucose hydrolysis and oxidation in supercritical water. *AIChE Journal* 41 (3):637–648.
38. Yu DH, Aihara M, Antal MJ (1993) Hydrogen-production by steam reforming glucose in supercritical water. *Energy & Fuels* 7 (5):574–577.
39. Antal MJ, Allen SG, Schulman D, Xu XD, Divilio RJ (2000) Biomass gasification in supercritical water. *Industrial & Engineering Chemistry Research* 39 (11):4040–4053.
40. Xu XD, Antal MJ (1998) Gasification of sewage sludge and other biomass for hydrogen production in supercritical water. *Environmental Progress* 17 (4):215–220.
41. Maiella PG, Brill TB (1996) Spectroscopy of hydrothermal reactions. 3. The water–gas reaction, "hot spots," and formation of volatile salts of NCO- from aqueous [$NH_3(CH_2)_nNH_3$]NO_3 (n=2,3) at 720 K and 276 bar by T-jump/FT-IR spectroscopy. *Applied Spectroscopy* 50 (7):829–835.
42. Marrone PA, Gschwend PM, Swallow KC, Peters WA, Tester JW (1998) Product distribution and reaction pathways for methylene chloride hydrolysis and oxidation under hydrothermal conditions. *The Journal of Supercritical Fluids* 12 (3):239–254.
43. Li L, Portela JR, Vallejo D, Gloyna EF (1999) Oxidation and hydrolysis of lactic acid in near-critical water. *Industrial & Engineering Chemistry Research* 38 (7):2599–2606.
44. Helling RK, Tester JW (1987) Oxidation kinetics of carbon monoxide in supercritical water. *Energy & Fuels* 1 (5):417–423.
45. Holgate HR, Tester JW (1994) Oxidation of hydrogen and carbon-monoxide in subcritical and supercritical water—Reaction-kinetics, pathways, and water-density effects. 1. Experimental results. *Journal of Physical Chemistry* 98 (3):800–809.
46. Holgate HR, Webley PA, Tester JW, Helling RK (1992) Carbon monoxide oxidation in supercritical water—The effects of heat-transfer and the water gas shift reaction on observed kinetics. *Energy & Fuels* 6 (5):586–597.
47. Adschiri T, Shibata R, Sato T, Watanabe M, Arai K (1998) Catalytic hydrodesulfurization of dibenzothiophene through partial oxidation and a water–gas shift reaction in supercritical water. *Industrial & Engineering Chemistry Research* 37 (7):2634–2638.
48. Arai K, Adschiri T, Watanabe M (2000) Hydrogenation of hydrocarbons through partial oxidation in supercritical water. *Industrial & Engineering Chemistry Research* 39 (12):4697–4701.
49. Matsumura Y, Nonaka H, Yokura H, Tsutsumi A, Yoshida K (1999) Co-liquefaction of coal and cellulose in supercritical water. *Fuel* 78 (9):1049–1056.
50. Shen Z, Zhang YL, Jin FM (2011) From $NaHCO_3$ into formate and from isopropanol into acetone: Hydrogen-transfer reduction of $NaHCO_3$ with isopropanol in high-temperature water. *Green Chemistry* 13 (4):820–823.
51. Shen Z, Zhang YL, Jin FM (2012) The alcohol-mediated reduction of CO_2 and $NaHCO_3$ into formate: A hydrogen transfer reduction of $NaHCO_3$ with glycerine under alkaline hydrothermal conditions. *RSC Adv* 2 (3):797–801.
52. Kruse A, Ebert KH (1996) Chemical reactions in supercritical water. 1. Pyrolysis of *tert*-butylbenzene. *Berichte der Bunsengesellschaft für physikalische Chemie* 100 (1):80–83.
53. Zhang YL, Shen Z, Zhou XF, Zhang M, Jin FM (2012) Solvent isotope effect and mechanism for the production of hydrogen and lactic acid from glycerol under hydrothermal alkaline conditions. *Green Chemistry* 14 (12):3285–3288.
54. Ramayya S, Brittain A, Dealmeida C, Mok W, Antal MJ (1987) Acid-catalyzed dehydration of alcohols in supercritical water. *Fuel* 66 (10):1364–1371.

55. Narayan R, Antal MJ (1989) Kinetic elucidation of the acid-catalyzed mechanism of 1-propanol dehydration in supercritical water. *ACS Symposium Series* 406:226–241.
56. Xu XD, Dealmeida CP, Antal MJ (1991) Mechanism and kinetics of the acid-catalyzed formation of ethene and diethyl-ether from ethanol in supercritical water. *Industrial & Engineering Chemistry Research* 30 (7):1478–1485.
57. Antal MJ, Leesomboon T, Mok WS, Richards GN (1991) Kinetic studies of the reactions of ketoses and aldoses in water at high-temperature. 3. Mechanism of formation of 2-furaldehyde from d-xylose. *Carbohydrate Research* 217:71–85.
58. Xu XD, Antal MJ (1994) Kinetics and mechanism of isobutene formation from *t*-butanol in hot liquid water. *AIChE Journal* 40 (9):1524–1534.
59. Xu XD, Antal MJ, Anderson DGM (1997) Mechanism and temperature-dependent kinetics of the dehydration of *tert*-butyl alcohol in hot compressed liquid water. *Industrial & Engineering Chemistry Research* 36 (1):23–41.
60. Antal MJ, Carlsson M, Xu X, Anderson DGM (1998) Mechanism and kinetics of the acid-catalyzed dehydration of 1- and 2-propanol in hot compressed liquid water. *Industrial & Engineering Chemistry Research* 37 (10):3820–3829.
61. Arita T, Nakahara K, Nagami K, Kajimoto O (2003) Hydrogen generation from ethanol in supercritical water without catalyst. *Tetrahedron Letters* 44 (5):1083–1086.
62. Shen Z, Jin FM, Zhang YL, Wu B, Cao JL (2009) Hydrogen transfer reduction of ketones using formic acid as a hydrogen donor under hydrothermal conditions. *Journal of Zhejiang University Science A* 10 (11):1631–1635.
63. Shen Z, Zhang YL, Jin FM, Zhou XF, Kishita A, Tohji K (2010) Hydrogen-transfer reduction of ketones into corresponding alcohols using formic acid as a hydrogen donor without a metal catalyst in high-temperature water. *Industrial & Engineering Chemistry Research* 49 (13):6255–6259.
64. Wang XG, Gron LU, Klein MT, Brill TB (1995) The influence of high-temperature water on the reaction pathways of nitroanilines. *The Journal of Supercritical Fluids* 8 (3):236–249.
65. Belsky AJ, Maiella PG, Brill TB (1999) Spectroscopy of hydrothermal reactions. 13. Kinetics and mechanisms of decarboxylation of acetic acid derivatives at 100–260°C under 275 bar. *Journal of Physical Chemistry A* 103 (21):4253–4260.
66. Takahashi H, Hisaoka S, Nitta T (2002) Ethanol oxidation reactions catalyzed by water molecules: $CH_3CH_2OH+nH_2O = CH_3CHO+H_2+nH_2O$ ($n = 0, 1, 2$). *Chemical Physics Letters* 363 (1–2):80–86.
67. Jin FM, Zhou ZY, Enomoto H, Moriya T, Higashijima H (2004) Conversion mechanism of cellulosic biomass to lactic acid in subcritical water and acid–base catalytic effect of subcritical water. *Chemistry Letters* 33 (2):126–127.
68. Chheda JN, Huber GW, Dumesic JA (2007) Liquid-phase catalytic processing of biomass-derived oxygenated hydrocarbons to fuels and chemicals. *Angewandte Chemie—International Edition* 46 (38):7164–7183.
69. Bobleter O (2005) Hydrothermal degradation and fractionation of saccharides and polysaccharides. In Dumitriu S (ed) *Polysaccharides: Structural Diversity and Functional Versatility*, 2nd edn. Marcel Dekker, New York, pp. 893–937.
70. Huang Y-B, Fu Y (2013) Hydrolysis of cellulose to glucose by solid acid catalysts. *Green Chemistry* 15 (5):1095–1111.
71. Sasaki M, Kabyemela B, Malaluan R, Hirose S, Takeda N, Adschiri T, Arai K (1998) Cellulose hydrolysis in subcritical and supercritical water. *The Journal of Supercritical Fluids* 13 (1–3):261–268.
72. Sasaki M, Fang Z, Fukushima Y, Adschiri T, Arai K (2000) Dissolution and hydrolysis of cellulose in subcritical and supercritical water. *Industrial & Engineering Chemistry Research* 39 (8):2883–2890.
73. Brunner G (2009) Near critical and supercritical water. Part I. Hydrolytic and hydrothermal processes. *The Journal of Supercritical Fluids* 47 (3):373–381.

74. Takagaki A, Nishimura S, Ebitani K (2012) Catalytic transformations of biomass-derived materials into value-added chemicals. *Catalysis Surveys from Asia* 16 (3):164–182.
75. Kabyemela BM, Adschiri T, Malaluan RM, Arai K (1999) Glucose and fructose decomposition in subcritical and supercritical water: Detailed reaction pathway, mechanisms, and kinetics. *Industrial & Engineering Chemistry Research* 38 (8):2888–2895.
76. Wang Y, Kovacik R, Meyer B, Kotsis K, Stodt D, Staemmler V, Qiu H, Traeger F, Langenberg D, Muhler M, Woell C (2007) CO_2 activation by ZnO through the formation of an unusual tridentate surface carbonate. *Angewandte Chemie—International Edition* 46 (29):5624–5627.
77. Kabyemela BM, Adschiri T, Malaluan R, Arai K (1997) Degradation kinetics of dihydroxyacetone and glyceraldehyde in subcritical and supercritical water. *Industrial & Engineering Chemistry Research* 36 (6):2025–2030.
78. Antal Jr MJ, Mok WSL, Richards GN (1990) Mechanism of formation of 5-(hydroxymethyl)-2-furaldehyde from d-fructose and sucrose. *Carbohydrate Research* 199 (1):91–109.
79. Amarasekara AS, Williams LD, Ebede CC (2008) Mechanism of the dehydration of d-fructose to 5-hydroxymethylfurfural in dimethyl sulfoxide at 150 degrees C: An NMR study. *Carbohydrate Research* 343 (18):3021–3024.
80. Guan J, Cao Q, Guo X, Mu X (2011) The mechanism of glucose conversion to 5-hydroxymethylfurfural catalyzed by metal chlorides in ionic liquid: A theoretical study. *Computational and Theoretical Chemistry* 963 (2–3):453–462.
81. Weingarten R, Conner WC, Huber GW (2012) Production of levulinic acid from cellulose by hydrothermal decomposition combined with aqueous phase dehydration with a solid acid catalyst. *Energy and Environmental Science* 5 (6):7559–7574.
82. Wettstein SG, Alonso DM, Chong YX, Dumesic JA (2012) Production of levulinic acid and gamma-valerolactone (GVL) from cellulose using GVL as a solvent in biphasic systems. *Energy and Environmental Science* 5 (8):8199–8203.
83. Bond JQ, Alonso DM, Wang D, West RM, Dumesic JA (2010) Integrated catalytic conversion of gamma-valerolactone to liquid alkenes for transportation fuels. *Science* 327 (5969):1110–1114.
84. Asghari FS, Yoshida H (2006) Acid-catalyzed production of 5-hydroxymethyl furfural from D-fructose in subcritical water. *Industrial & Engineering Chemistry Research* 45 (7):2163–2173.
85. Srokol Z, Bouche AG, van Estrik A, Strik RCJ, Maschmeyer T, Peters JA (2004) Hydrothermal upgrading of biomass to biofuel; studies on some monosaccharide model compounds. *Carbohydrate Research* 339 (10):1717–1726.
86. Takeuchi Y, Jin FM, Tohji K, Enomoto H (2008) Acid catalytic hydrothermal conversion of carbohydrate biomass into useful substances. *Journal of Materials Science* 43 (7):2472–2475.
87. Shen J, Wyman CE (2012) Hydrochloric acid-catalyzed levulinic acid formation from cellulose: Data and kinetic model to maximize yields. *AIChE Journal* 58 (1):236–246.
88. Cinlar B, Wang TF, Shanks BH (2013) Kinetics of monosaccharide conversion in the presence of homogeneous Bronsted acids. *Applied Catalysis A—General* 450:237–242.
89. Weingarten R, Cho J, Xing R, Conner WC, Huber GW (2012) Kinetics and reaction engineering of levulinic acid production from aqueous glucose solutions. *Chemsuschem* 5 (7):1280–1290.
90. Pagan-Torres YJ, Wang TF, Gallo JMR, Shanks BH, Dumesic JA (2012) Production of 5-hydroxymethylfurfural from glucose using a combination of lewis and bronsted acid catalysts in water in a biphasic reactor with an alkylphenol solvent. *ACS Catal* 2 (6):930–934.

91. Yang Y, Hu CW, Abu-Omar MM (2012) Conversion of carbohydrates and lignocellulosic biomass into 5-hydroxymethylfurfural using $AlCl_3·6H_2O$ catalyst in a biphasic solvent system. *Green Chemistry* 14 (2):509–513.
92. Wang TF, Pagan-Torres YJ, Combs EJ, Dumesic JA, Shanks BH (2012) Water-compatible Lewis acid-catalyzed conversion of carbohydrates to 5-hydroxymethylfurfural in a biphasic solvent system. *Topics in Catalysis* 55 (7–10):657–662.
93. Roman-Leshkov Y, Chheda JN, Dumesic JA (2006) Phase modifiers promote efficient production of hydroxymethylfurfural from fructose. *Science* 312 (5782):1933–1937.
94. Kishida H, Jin FM, Yan XY, Moriya T, Enomoto H (2006) Formation of lactic acid from glycolaldehyde by alkaline hydrothermal reaction. *Carbohydrate Research* 341 (15):2619–2623.
95. Sasaki M, Goto K, Tajima K, Adschiri T, Arai K (2002) Rapid and selective retro-aldol condensation of glucose to glycolaldehyde in supercritical water. *Green Chemistry* 4 (3):285–287.
96. Yang BY, Montgomery R (1996) Alkaline degradation of glucose: Effect of initial concentration of reactants. *Carbohydrate Research* 280 (1):27–45.
97. Yan X, Jin F, Tohji K, Kishita A, Enomoto H (2010) Hydrothermal conversion of carbohydrate biomass to lactic acid. *AIChE Journal* 56 (10):2727–2733.
98. Yan XY, Jin FM, Tohji K, Moriya T, Enomoto H (2007) Production of lactic acid from glucose by alkaline hydrothermal reaction. *Journal of Materials Science* 42 (24):9995–9999.
99. Sánchez C, Egüés I, García A, Llano-Ponte R, Labidi J (2012) Lactic acid production by alkaline hydrothermal treatment of corn cobs. *Chemical Engineering Journal* 181–182:655–660.
100. Jin FM, Yun J, Li GM, Kishita A, Tohji K, Enomoto H (2008) Hydrothermal conversion of carbohydrate biomass into formic acid at mild temperatures. *Green Chemistry* 10 (6):612–615.
101. Yu JL, Savage PE (1998) Decomposition of formic acid under hydrothermal conditions. *Industrial & Engineering Chemistry Research* 37 (1):2–10.
102. Akiya N, Savage PE (1998) Role of water in formic acid decomposition. *AIChE Journal* 44 (2):405–415.
103. Wakai C, Yoshida K, Tsujino Y, Matubayasi N, Nakahara M (2004) Effect of concentration, acid, temperature, and metal on competitive reaction pathways for decarbonylation and decarboxylation of formic acid in hot water. *Chemistry Letters* 33 (5):572–573.
104. Liu J, Zeng X, Cheng M, Yun J, Li Q, Jing Z, Jin F (2012) Reduction of formic acid to methanol under hydrothermal conditions in the presence of Cu and Zn. *Bioresource Technology* 114:658–662.
105. Wang W, Wang SP, Ma XB, Gong JL (2011) Recent advances in catalytic hydrogenation of carbon dioxide. *Chemical Society Reviews* 40 (7):3703–3727.
106. Saito K, Kakumoto T, Kuroda H, Torii S, Imamura A (1984) Thermal unimolecular decomposition of formic acid *Journal of Chemical Physics* 80 (10):4989–4997.

2 Catalytic Hydrothermal Reactions for Small Molecules Activation

Yuanqing Wang and Fangming Jin

CONTENTS

2.1 General Considerations...23
    2.1.1 Interfacial Chemistry between Solid and Solution.........................23
    2.1.2 Kinetic Model...24
    2.1.3 *In-situ* Characterization..25
    2.1.4 Quantum Chemistry Calculation..26
    2.1.5 Acidic/Basic and Redox Catalysis...26
2.2 Case Study 1: Hydrothermal Reactions of Aqueous Sulfur Species..............27
2.3 Case Study 2: Conversion of CO_2 into Organic Acid....................................29
2.4 Conclusions and Outlook..31
References...32

2.1 GENERAL CONSIDERATIONS

2.1.1 INTERFACIAL CHEMISTRY BETWEEN SOLID AND SOLUTION

The interfacial chemistry between solid and solution is always discussed in the field of electrochemistry [1,2]. Therefore, the solid discussed is an electrode, and the solution is an electrolyte solution. In this chapter, we restrict the theme to the interfacial chemistry between electrode and electrolyte solution, which can be regarded as a main part covering most solid catalysts used (metal and semiconductor) and solutions containing ions. Readers are recommended to read the above references for detailed explanation. Here, some key points are introduced.

When a solid (electrode) is immersed in an aqueous electrolyte, several processes at the interface occur. First of all, the ions in the electrolyte may be adsorbed at the surface, which can be divided into nonspecific adsorption and specific adsorption based on the interaction with the solid. Second, solvent (water) may also be adsorbed at the surface. These phenomena, including adsorption and the presence of ions near the surface, would induce surface charging of solid to neutralize the charge in the solution side. Consequently, there is an equal and opposite charge in the solid and solution interface that is therefore called electrical double layer. The solutions side is thought to be composed of many layers. The adsorption of ions with a weak solvation shell that form a chemical bond to the solid surface is termed "specific." The center of specifically adsorbed anions and cations is defined as the inner Helmholtz plane. On the contrary, if the interaction of the solvated ions with the charged solid includes only long-range electrostatic forces, the adsorption, essentially independent of the chemical properties of the ions, is considered to be nonspecific adsorption. The center of these nearest ions to the surface is defined as the outer Helmholtz plane. These ions are distributed in the region called diffuse layer extending from the outer Helmholtz plane to the bulk solution.

The property of electrical double layer can therefore affect the reaction at the interface. For instance, Chen et al. [3] found that the cation-induced local field in the electrical double layer would stabilize the adsorbed molecule and thus shift the reaction free energy profile based on density functional theory calculation, because the energies of adsorbed molecules possessing large dipoles or polarizabilities would be largely affected by the induced electric field in the interface. Although these discussions are presented in the context of electrochemical reaction with an external bias, the importance of electrical double layer should be addressed if reaction occurs at the solid/solution interface.

2.1.2 Kinetic Model

There are mainly three types of reaction mechanism proposed to describe the kinetics of reaction at the surface. The mechanism is originally developed for the solid/gas interface reaction. One is called the Langmuir–Hinshelwood mechanism, which states that the products are from the reaction of two adsorbed species as shown in Equations 2.1 through 2.4.

$$A + * \Leftrightarrow A* \tag{2.1}$$

$$B + * \Leftrightarrow B* \tag{2.2}$$

$$A* + B* \Leftrightarrow AB* + * \tag{2.3}$$

$$AB* \Leftrightarrow AB + *, \tag{2.4}$$

where A and B denote reactants and AB denotes product; * denotes surface sites available for adsorption; A* or B* or AB* denotes the adsorbed species (intermediates). Equations 2.1 through 2.4 describe the simplest case in this mechanism. Things can become complicated if a dissociative adsorption occurs instead of associative adsorption. There are probably more than one elementary step for Equation 2.3.

The second mechanism is the Eley–Rideal mechanism, which states that the reactant (not adsorbed) from the gas phase reacts with the adsorbed species as shown in Equation 2.5.

$$A + B* \Leftrightarrow AB* \Leftrightarrow AB + * \tag{2.5}$$

The third mechanism, originated from selective oxidation catalysis, is the Mars–van Krevelen mechanism [4], which states that the whole reaction can be divided into two separate redox reactions as shown in Equations 2.6 and 2.7.

Step 1. Aromatic compound + oxidized catalyst
$$\rightarrow \text{oxidation products} + \text{reduced catalyst} \tag{2.6}$$

$$\text{Step 2. Reduced catalyst} + \text{oxygen} \rightarrow \text{oxidized catalyst} \tag{2.7}$$

In step 2, the molecular oxygen is proposed to follow the transformation on oxide catalysts [5] as shown in Equation 2.8,

$$O_2(g) \xrightarrow{\text{adsorption}} O_2(s) \xrightarrow{+e^-} O_2^-(s) \xrightarrow{+e^-} O_2^{2-}(s) \xrightarrow{\text{dissociation}}$$
$$2O^-(s) \xrightarrow{+2e^-} 2O^{2-}(lc), \tag{2.8}$$

where g represents gas phase, s represents surface, and lc represents lattice. This transformation would produce two types of oxygen species: an electrophilic one and a nucleophilic one. The former, which includes O^{2-} and O^- species, can interact with pi electrons and make C–C bonds cleaved, which finally leads to total oxidation with CO and CO_2 as the products. The latter, which includes O^{2-}, can facilitate hydrogen abstraction and oxygen insertion, which leads to selective oxidation. Besides, the lattice oxygen diffusion between the bulk phase and surface may also affect activity and selectivity.

Different mechanisms may show different kinetic behavior. The conventional way is to fit the kinetic data with the set model using quasi-equilibrium approximation or steady-state approximation. However, the agreement between experiments and the set model cannot always guarantee the existence of the mechanism behind the set model. Therefore, *in-situ* characterization of the catalysts or system for determining the properties of catalysts or intermediates is always helpful to unravel the complicated mechanism.

2.1.3 IN-SITU CHARACTERIZATION

In-situ characterization of the hydrothermal reaction is not easy since the reaction is conducted at high-temperature and high-pressure conditions. To maintain the closed cell at such conditions and be available for monitoring, pressure-resistant and transparent material is required, such as diamond. One apparatus to fulfill these requirements is the diamond anvil cell (DAC) [6]. Analytical measurement reportedly using DAC includes optical microscopy [7], Raman [8], and infrared spectra [9]. Optical microscopy observation on the dissolution of cellulose particle at high temperature and high pressure revealed that the process is dependent on water density [7]. At a water density from 600 to 800 kg/m³, cellulose particles swelled noticeably and the temperature at which cellulose dissolved decreased with the increase in water density. At a water density greater than 900 kg/m³, the dissolution temperature increased. An *in-situ* Raman spectra study on the hydrothermal reactions of benzothiophene indicates that the desulfurization undergoes two steps, including hydrogenation and cleavage of the benzene ring–S bond and the C2–S bond [8]. It is worth noting that the Raman spectroscopic tool is very suitable for the hydrothermal system because water molecule vibration is not sensitive to Raman compared with infrared spectra. An infrared spectroscopic tool on the aqueous NaCl solution at temperatures from 25°C to 850°C and pressures up to 3 GPa was adopted to monitor the phase transition [9]. However, because of the expensive cell used, the numbers of reports relating to *in-situ* characterization are relatively limited. In addition, the analysis of spectral results is complicated because the chemistry under hydrothermal conditions may be different from that under ambient conditions and the peak or band position may shift according to a specific chemical environment. To overcome this, quantum chemistry calculation is a powerful tool to predict the chemical properties of studied system as well as the reaction pathway and intermediates.

2.1.4 Quantum Chemistry Calculation

As mentioned above, quantum chemistry calculation has now become an indispensable tool in catalysis research. Generally, reaction energies, including transition-state calculations, can predict the most thermodynamically or kinetically favorable reaction pathway in a set system. The population analysis or molecular orbital drawings can explain the reactivity in a quantitative way. Frequency or phonon calculations can predict the infrared and Raman spectra and can point out the vibration mode, which would be difficult to obtain from experimental results. Excitation energy calculations can provide information on the excitation of molecules, which can be related to UV–Vis spectra. In solid calculations, the determination of band structure and density of state could explain the catalysis mechanism of a solid catalyst. As in the field of hydrothermal reaction, many publications [10–14] have emerged recently to help understand the reactions at a molecular level. Akiya and Savage explored the effect of water on formic acid decomposition under hydrothermal conditions [10,15]. They calculated the decarboxylation and dehydration of formic acid and the isomerization of formic acid in the presence of water and found that water is a homogeneous catalyst for the former two reactions and is independent for the isomerization. A cluster model containing five Zn atoms and water molecules was recently set up to simulate the hydrothermal reactions of metal Zn within the mixed implicit–explicit solvation model [12,13]. The exhaustive search for the lowest-energy species finally indicates that the reduction of CO_2 by Zn undergoes via a Zn–H intermediate through an SN2-like mechanism.

2.1.5 Acidic/Basic and Redox Catalysis

Here, we take the hydrothermal reactions of biomass (cellulose) [16–21] as examples to explain the catalysis mechanism under hydrothermal conditions. C6 sugars can selectively produce 5-hydroxymethyl-2-furaldehyde (HMF), an important chemical intermediate, via homogeneous acids (H_3PO_4, H_2SO_4, or HCl) [22]. The selectivity for producing HMF using the three acids followed the following order: $H_3PO_4 > H_2SO_4 >$ HCl. The rehydration of HMF with two molecules of water by solid acid catalysts (ZrP and Amberlyst 70) would produce levulinic acid and formic acid [23]. Levulinic acid can be further converted into gamma-valerolactone (GVL) via hydrogenation by a Ru–Sn/C catalyst [24]. The produced GVL can be converted into liquid alkenes by SiO_2/Al_2O_3 and HZSM-5 or Amberlyst 70 [25].

The studies on the base-catalyzed reactions of glucose and cellulose at ~300°C show that fructose and some C3 carbon compounds, such as glyceraldehyde, dihydroxyacetone, and pyruvaldehyde, are selectively formed [26–30]. It was also found that isomerization occurred between glyceraldehyde and dihydroxyacetone followed by the subsequent dehydration to pyruvaldehyde. Transformation of pyruvaldehyde can lead to lactic acid, while the intermediates glycoaldehyede and erythrose can also produce lactic acid. Yan et al. [31,32] found that base catalysts NaOH and Ca(OH)$_2$ can increase the yield of lactic acid from carbohydrate. The yield of lactic acid from glucose was 27% with 2.5 M NaOH and 20% with 0.32 M Ca(OH)$_2$ under hydrothermal conditions. Further improvement of lactic acid yield (59% from glucose and 13.8% from cellulose) can be obtained using NaOH and the oxidant CuO [33,34]. Furthermore, the adoption of CuO and O$_2$ enables CuO as a catalyst in the whole catalytic cycle and O$_2$ as an oxidant [35]. The above two results indicate that the catalytic mechanism may obey the Mars–van Krevelen mechanism. Gao et al. for the first time used bentonite as a solid base catalyst to increase the yield of lactic acid from 5% (without bentonite) to 11% (with bentonite) from glucose [36].

2.2 CASE STUDY 1: HYDROTHERMAL REACTIONS OF AQUEOUS SULFUR SPECIES

Hydrothermal reactions of small molecules, such as CO_2, H_2S, and NO_3^-, have attracted much attention because of their implications in geochemistry [37,38], energy industry [39,40], and corrosion science [41,42]. The presence of metals and metal sulfides will significantly affect the reaction. Setiani et al. [43,44] and Wang et al. [45] studied the hydrothermal reaction of sulfide for hydrogen production as shown in Equation 2.9.

$$HS^- + H_2O \rightarrow S_xO_y^{2-} + H_2\uparrow \tag{2.9}$$

The oxidation products including $S_2O_3^{2-}$, $S_3O_6^{2-}$, SO_3^{2-}, and SO_4^{2-} have been detected by capillary electrophoresis [43–45]. Wang et al. [46] proved the catalytic role of Ni_3S_2 for the hydrogen production. As shown in Figure 2.1a, it was proposed that Ni_3S_2 may act as a semiconductor catalyst to receive the electrons from HS$^-$ and donate them to water, which therefore leads to hydrogen production. A key assumption is that under alkaline hydrothermal conditions, the SO_4^{2-}/S^{2-} redox couple may increase the energy level to the appropriate position above the conduction band edge position of Ni_3S_2 and donates electrons to Ni_3S_2. Interestingly, another group from the field of geochemistry [47] also proposed an electron flow in the conductive chimney that is composed of metal sulfides functioning as a catalyst based on geological findings and electrochemical measurements. As shown in Figure 2.1b, the S^0/S^{2-} redox couple in the hydrothermal vent may generate electrons in the inner surface of the chimney and the electrons are transferred to the outer surface of the chimney in which cathodic reaction occurs. It was later found that the cathodic reaction could be the reduction of CO_2 by nickel-containing iron sulfide [48] and nitrate reduction by molybdenum sulfide [49] in addition to noticeable hydrogen evolution. Therefore, the

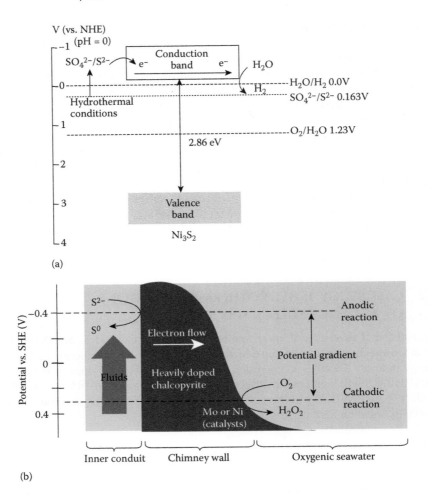

FIGURE 2.1 (a) Proposed mechanism of Ni_3S_2-catalyzed hydrogen production from water by HS^-. (Reprinted from Wang, Y., Jin, F., Zeng, X. et al., *Appl. Energ.*, 104, 306–309, 2013.) (b) Schematic illustration of an energy diagram for the anodic and cathodic reactions occurring at the inner and outer surfaces of the chimney. (Reprinted from Nakamura, R., Takashima, T., Kato, S. et al., *Angew. Chem. Int. Ed.*, 49, 7692–7694, 2010.)

catalytic role of metal sulfides under hydrothermal conditions has been separately investigated and generalized as a semiconductor catalyst. Further attention should be paid to the more detailed band structure of the semiconductor catalyst in working condition because adsorption may lead to band bending at the interface, which is directly related to catalysis [50].

Later, Wang et al. systematically investigated the hydrothermal reactions of sulfur species (HS^-, S, and $S_2O_3^{2-}$) in the presence of metals and metal sulfide [45]. The results show that there are mainly three types of reactions among them: (i) sulfidation of metals and metal sulfide by sulfur species, (ii) disproportionation of sulfur

Catalytic Hydrothermal Reactions for Small Molecules Activation

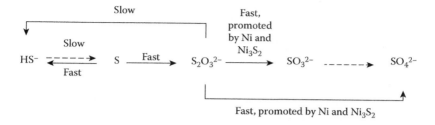

SCHEME 2.1 Proposed reaction pathways of sulfur species with water and metal or mineral under hydrothermal conditions (the solid lines represent confirmed reactions; the dotted lines represent unconfirmed reactions). (Reprinted from Wang, Y., Wang, F., Jin, F. et al., *Ind. Eng. Chem. Res.*, 52, 5616–5625, 2013.)

species, and (iii) oxidation of sulfur species by water to produce hydrogen. As shown in Scheme 2.1, HS^- is very stable with respect to the oxidation at 250°C even in the presence of metals and metal sulfide. Elemental sulfur would disproportionate actively to HS^- and $S_2O_3^{2-}$. In addition to disproportion, elemental sulfur would also be oxidized by water to form $S_2O_3^{2-}$ and H_2 in the presence of Fe and Ni_3S_2. $S_2O_3^{2-}$ was fast consumed with the formation of SO_3^{2-} and SO_4^{2-} in the presence of Ni and Ni_3S_2, whereas $S_2O_3^{2-}$ is relatively stable whether Fe was added or not.

2.3 CASE STUDY 2: CONVERSION OF CO_2 INTO ORGANIC ACID

The weakly alkaline hydrothermal reactions of bicarbonate and glycerol (or polyols) were developed by Jin's group [51,52] to produce formate and lactate that can be acidified further to organic acids. Bicarbonate, as a hydrated form of CO_2 according to the pH, is not differentiated with CO_2 strictly here. Glycerol, as the feedstock and reductant, can be obtained from the by-product of producing biodiesel mixed with water and other impurities. One advantage of this process is that the obtained glycerol can be directly introduced into the reactor without further purification. The reduction of bicarbonate into formate at the same time enables the utilization of CO_2 as a C1 building block for fuels and chemicals. Hydrogen transfer from glycerol to hydrogenate CO_2 or bicarbonate was proposed as a possible reaction mechanism in these studies [51–54]. In this section, the strategy of combining experiments (kinetic model) and quantum chemistry calculation is presented to shed light on the mechanistic insights.

Considering the present experimental findings [11,30,52,55–57], a simple kinetic model is set up and drawn in Scheme 2.2. As shown in Scheme 2.2, there are two parallel reaction pathways for conversion of glycerol by base catalyst. One is the main reaction that represents the simultaneous reduction of bicarbonate into formate and glycerol transformation to lactate. The other is a side reaction that also cannot be ignored. The reactions are then assumed to be first order with respect to the substrate. The initial concentration of bicarbonate is five times that of glycerol and can be viewed as a constant incorporated into the rate constant. According to these assumptions, the variation of glycerol conversion and product yield with reaction time can be expressed by Equations 2.10 through 2.12.

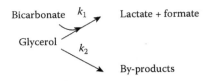

SCHEME 2.2 Kinetic model of glycerol conversion. (Reprinted from Wang, Y., Wang, F., Li, C. et al., *Int. J. Hydrogen Energy*, 41, 9128–9134, 2016.)

$$\frac{d[Gly]}{dt} = -(k_1 + k_2)[Gly] \quad (2.10)$$

$$\frac{d[LA]}{dt} = k_1[Gly] \quad (2.11)$$

$$\frac{d[FA]}{dt} = k_1[Gly], \quad (2.12)$$

where [Gly], [LA], and [FA] represent the concentration of glycerol, lactate, and formate, respectively; k_1 represents the rate constant for the main reaction while k_2 represents that for the side reaction. The experimental data were then used to fit these kinetic equations, and the fitted parameters of the Arrhenius equation are given in Table 2.1. As can be seen from Table 2.1, the apparent activation energy of the main reaction is 41 kcal/mol and that of the side reaction is 33 kcal/mol. These estimated values are comparable to the reported values of glycerol conversion in near-critical water, which are 36 kcal/mol [58] and 33 kcal/mol [59].

Further quantum chemistry calculations reveal that the key step in the whole process [51–53], from hydroxyacetone and bicarbonate to pyruvaldehyde and formate, is composed of two elementary steps (see Figure 2.2). The first step (step 1) is the deprotonation of the –OH group in hydroxyacetone by bicarbonate. This step would produce $CH_3COCH_2O^-$ and release CO_2 molecules. The second step (step 2) proceeds in a way that one hydrogen in the –CH_2O^- group attacks the carbon on CO_2 by nucleophilic addition to form pyruvaldehyde and formate. $CH_3COCH_2O^-$ is first suggested by theoretical calculation to be the key intermediate of reducing CO_2. The role of bicarbonate can be concluded into two aspects: (1) to deprotonate the –OH group in hydroxyacetone as a base and (2) to release CO_2 molecules to be reduced in the second step. As can be seen from Figure 2.2, the estimated activation barriers

TABLE 2.1
Fitted Arrhenius Equation Parameters

Rate Constant	Fitted k Values (min^{-1})			ln A (min^{-1})	E_a (kcal/mol)
	275°C	300°C	325°C		
k_1	0.0017 ± 0.0002	0.0089 ± 0.0005	0.0411 ± 0.0049	31.7	41
k_2	0.0018 ± 0.0004	0.0033 ± 0.0009	0.0233 ± 0.0075	23.8	33

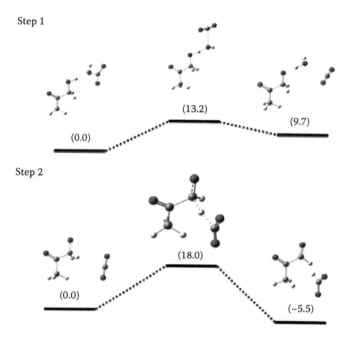

FIGURE 2.2 Free energy diagram of the two-step mechanism from hydroxyacetone and bicarbonate to pyruvaldehyde and formate. The values in parentheses are Gibbs free energy (kcal/mol) calculated at 573 K. (Reprinted from Wang, Y., Wang, F., Li, C. et al., *Int. J. Hydrogen Energy*, 41, 9128–9134, 2016.)

(13.2 and 18.0 kcal/mol) are easy to overcome under hydrothermal conditions and are quite different from the apparent activation energy (41 kcal/mol). This result suggests that the rate-determining step of this process is some step before hydroxyacetone reduction or after pyruvaldehyde formation. Therefore, the optimal catalyst (base) should be searched to accelerate the conversion of glycerol to hydroxyacetone or pyruvaldehyde to lactate. Principally, the basicity of medium is critical to determine the selectivity. If the base is too strong, another pathway will be favored more [52]. If the base is too weak, hydroxyacetone or lactate may not be formed readily.

2.4 CONCLUSIONS AND OUTLOOK

This chapter first introduces the general considerations of catalysis under hydrothermal conditions from key concepts to characterization tools and methods with the examples mainly conducted in Jin's group [60]. Then, two case studies including heterogeneous and homogeneous reactions from the author are reviewed. However, the molecular understanding of hydrothermal reactions of small molecules is still limited compared with its counterparts, which are thermocatalysis at the solid/gas interface and electrocatalysis at the solid/solution interface. The reasons may include but are not limited to the following: (1) the insufficiency of application of *in-situ* characterization studies on the hydrothermal reactions for determination of intermediates or solid surface chemistry and (2) the lack of support from quantum chemistry calculation tools that are normally coded for reactions under non-extreme conditions (unlike hydrothermal conditions) or the shortage of computational accuracy with reasonable costs for large or complex systems. Besides, it is still under debate as to what extent the concepts of catalysis (from thermocatalysis and electrocatalysis) can be applied in hydrothermal reactions. High-temperature water exhibits quite unique properties that are different from those of ambient liquid water, such as lower dielectric constant, fewer and weaker hydrogen bonding, lower density and viscosity, and higher ion product (K_w). All these properties may lead to unique reactivity under hydrothermal conditions. Hopefully, the advances in experimental tools and theories would answer these questions and reveal new concepts of catalysis under hydrothermal conditions.

REFERENCES

1. Bard, A. and Faulkner, L., *Electrochemical Methods: Fundamentals and Applications*. New York: John Wiley & Sons, 2001.
2. Strasser, P. and Ogasawara, H., Chapter 6—Surface Electrochemistry A2—Nilsson, Anders. In *Chemical Bonding at Surfaces and Interfaces*, Pettersson, L. G. M.; Nørskov, J. K., Eds. Amsterdam: Elsevier 2008; pp. 397–455.
3. Chen, L. D., Urushihara, M., Chan, K. and Nørskov, J. K. Electric field effects in electrochemical CO_2 reduction. *ACS Catal* 2016: 7133–7139.
4. Mars, P. and van Krevelen, D. W. The proceedings of the conference on oxidation processes oxidations carried out by means of vanadium oxide catalysts. *Chemical Engineering Science* 1954; 3: 41–59.
5. Haber, J. and Turek, W. Kinetic studies as a method to differentiate between oxygen species involved in the oxidation of propene. *Journal of Catalysis* 2000; 190: 320–326.
6. Smith Jr, R. L. and Fang, Z. Techniques, applications and future prospects of diamond anvil cells for studying supercritical water systems. *The Journal of Supercritical Fluids* 2009; 47: 431–446.
7. Ogihara, Y., Smith, R. L., Inomata, H. and Arai, K. Direct observation of cellulose dissolution in subcritical and supercritical water over a wide range of water densities (550–1000 kg/m³). *Cellulose* 2005; 12: 595–606.
8. Huo, Z., Jin, F., Yao, G., Enomoto, H. and Kishita, A. An *in-situ* Raman spectroscopic study of benzothiophene and its desulfurization under alkaline hydrothermal conditions. *Industrial & Engineering Chemistry Research* 2015; 54: 1397–1406.
9. Zhang, R. and Hu, S. Hydrothermal study using a new diamond anvil cell with in situ IR spectroscopy under high temperatures and high pressures. *The Journal of Supercritical Fluids* 2004; 29: 185–202.
10. Akiya, N. and Savage, P. E. Role of water in formic acid decomposition. *AIChE Journal* 1998; 44: 405–415.
11. Wang, Y., Wang, F., Li, C. and Jin, F. Kinetics and mechanism of reduction of CO_2 by glycerol under alkaline hydrothermal conditions. *International Journal of Hydrogen Energy* 2016; 41: 9128–9134.
12. Ogata, K., Hatakeyama, M., Jin, F., Zeng, X., Wang, Y., Fujii, K. and Nakamura, S. A model study of hydrothermal reactions of trigonal dipyramidal Zn_5 cluster with two water molecules. *Computational and Theoretical Chemistry* 2015; 1070: 126–131.
13. Zeng, X., Hatakeyama, M., Ogata, K., Liu, J., Wang, Y., Gao, Q., Fujii, K., Fujihira, M., Jin, F. and Nakamura, S. New insights into highly efficient reduction of CO_2 to formic acid by using zinc under mild hydrothermal conditions: A joint experimental and theoretical study. *Physical Chemistry Chemical Physics* 2014; 16: 19836–19840.

14. Chia, M., Haider, M. A., Pollock, G., Kraus, G. A., Neurock, M. and Dumesic, J. A. Mechanistic insights into ring-opening and decarboxylation of 2-pyrones in liquid water and tetrahydrofuran. *Journal of the American Chemical Society* 2013; 135: 5699–5708.
15. Akiya, N. and Savage, P. E. Roles of water for chemical reactions in high-temperature water. *Chemical Reviews* 2002; 102: 2725–2750.
16. Tanksale, A., Beltramini, J. N. and Lu, G. M. A review of catalytic hydrogen production processes from biomass. *Renewable & Sustainable Energy Reviews* 2010; 14: 166–182.
17. Huber, G. W., Iborra, S. and Corma, A. Synthesis of transportation fuels from biomass: Chemistry, catalysts, and engineering. *Chemical Reviews* 2006; 106: 4044–4098.
18. Baiker, A. Supercritical fluids in heterogeneous catalysis. *Chemical Reviews* 1999; 99: 453–473.
19. Huang, Y.-B. and Fu, Y. Hydrolysis of cellulose to glucose by solid acid catalysts. *Green Chemistry* 2013; 15: 1095–1111.
20. Kobayashi, H., Komanoya, T., Guha, S. K., Hara, K. and Fukuoka, A. Conversion of cellulose into renewable chemicals by supported metal catalysis. *Applied Catalysis A—General* 2011; 409: 13–20.
21. Guo, Y., Wang, S. Z., Xu, D. H., Gong, Y. M., Ma, H. H. and Tang, X. Y. Review of catalytic supercritical water gasification for hydrogen production from biomass. *Renewable & Sustainable Energy Reviews* 2010; 14: 334–343.
22. Asghari, F. S. and Yoshida, H. Acid-catalyzed production of 5-hydroxymethyl furfural from D-fructose in subcritical water. *Industrial & Engineering Chemistry Research* 2006; 45: 2163–2173.
23. Weingarten, R., Conner, W. C. and Huber, G. W. Production of levulinic acid from cellulose by hydrothermal decomposition combined with aqueous phase dehydration with a solid acid catalyst. *Energy and Environmental Science* 2012; 5: 7559–7574.
24. Wettstein, S. G., Alonso, D. M., Chong, Y. X. and Dumesic, J. A. Production of levulinic acid and gamma-valerolactone (GVL) from cellulose using GVL as a solvent in biphasic systems. *Energy and Environmental Science* 2012; 5: 8199–8203.
25. Bond, J. Q., Alonso, D. M., Wang, D., West, R. M. and Dumesic, J. A. Integrated catalytic conversion of gamma-valerolactone to liquid alkenes for transportation fuels. *Science* 2010; 327: 1110–1114.
26. Jin, F. M., Zhou, Z. Y., Enomoto, H., Moriya, T. and Higashijima, H. Conversion mechanism of cellulosic biomass to lactic acid in subcritical water and acid-base catalytic effect of subcritical water. *Chemistry Letters* 2004; 33: 126–127.
27. Kabyemela, B. M., Adschiri, T., Malaluan, R. M. and Arai, K. Glucose and fructose decomposition in subcritical and supercritical water: Detailed reaction pathway, mechanisms, and kinetics. *Industrial & Engineering Chemistry Research* 1999; 38: 2888–2895.
28. Sasaki, M., Kabyemela, B., Malaluan, R., Hirose, S., Takeda, N., Adschiri, T. and Arai, K. Cellulose hydrolysis in subcritical and supercritical water. *The Journal of Supercritical Fluids* 1998; 13: 261–268.
29. Kabyemela, B. M., Adschiri, T., Malaluan, R. and Arai, K. Degradation kinetics of dihydroxyacetone and glyceraldehyde in subcritical and supercritical water. *Industrial & Engineering Chemistry Research* 1997; 36: 2025–2030.
30. Kishida, H., Jin, F. M., Yan, X. Y., Moriya, T. and Enomoto, H. Formation of lactic acid from glycolaldehyde by alkaline hydrothermal reaction. *Carbohydrate Research* 2006; 341: 2619–2623.
31. Yan, X., Jin, F., Tohji, K., Kishita, A. and Enomoto, H. Hydrothermal conversion of carbohydrate biomass to lactic acid. *AIChE Journal* 2010; 56: 2727–2733.
32. Yan, X. Y., Jin, F. M., Tohji, K., Moriya, T. and Enomoto, H. Production of lactic acid from glucose by alkaline hydrothermal reaction. *Journal of Materials Science* 2007; 42: 9995–9999.

33. Wang, F., Wang, Y., Jin, F., Yao, G., Huo, Z., Zeng, X. and Jing, Z. One-pot hydrothermal conversion of cellulose into organic acids with CuO as an oxidant. *Industrial & Engineering Chemistry Research* 2014; 53: 7939–7946.
34. Wang, Y., Jin, F., Sasaki, M., Wahyudiono, Wang, F., Jing, Z. and Goto, M. Selective conversion of glucose into lactic acid and acetic acid with copper oxide under hydrothermal conditions. *AIChE Journal* 2013; 59: 2096–2104.
35. Gao, X., Chen, X., Zhang, J., Guo, W., Jin, F. and Yan, N. Transformation of chitin and waste shrimp shells into acetic acid and pyrrole. *ACS Sustainable Chemistry & Engineering* 2016; 4: 3912–3920.
36. Gao, X., Zhong, H., Yao, G., Guo, W. and Jin, F. Hydrothermal conversion of glucose into organic acids with bentonite as a solid-base catalyst. *Catalysis Today* 2016; 274: 49–54.
37. Martin, W. and Russell, M. J. On the origin of biochemistry at an alkaline hydrothermal vent. *Philosophical Transactions of the Royal Society B—Biological Sciences* 2007; 362: 1887–1925.
38. Wachtershauser, G. On the chemistry and evolution of the pioneer organism. *Chemistry & Biodiversity* 2007; 4: 584–602.
39. Sato, T., Mori, S., Watanabe, M., Sasaki, M. and Itoh, N. Upgrading of bitumen with formic acid in supercritical water. *The Journal of Supercritical Fluids* 2010; 55: 232–240.
40. Sato, T., Trung, P. H., Tomita, T. and Itoh, N. Effect of water density and air pressure on partial oxidation of bitumen in supercritical water. *Fuel* 2012; 95: 347–351.
41. Marrone, P. A. and Hong, G. T. Corrosion control methods in supercritical water oxidation and gasification processes. *The Journal of Supercritical Fluids* 2009; 51: 83–103.
42. Kritzer, P. Corrosion in high-temperature and supercritical water and aqueous solutions: A review. *The Journal of Supercritical Fluids* 2004; 29: 1–29.
43. Setiani, P., Watanabe, N., Kishita, A. and Tsuchiya, N. Temperature- and pH-dependent mechanism of hydrogen production from hydrothermal reactions of sulfide. *Int J Hydrogen Energy* 2012; 37: 18679–18687.
44. Setiani, P., Watanabe, N., Kishita, A. and Tsuchiya, N. A predictive model and mechanisms associated with hydrogen production via hydrothermal reactions of sulfide. *Int J Hydrogen Energy* 2013; 38: 11209–11221.
45. Wang, Y., Wang, F., Jin, F. and Jing, Z. Effects of metals and Ni_3S_2 on reactions of sulfur species (HS^-, S, and $S_2O_3^{2-}$) under alkaline hydrothermal conditions. *Industrial & Engineering Chemistry Research* 2013; 52: 5616–5625.
46. Wang, Y., Jin, F., Zeng, X., Ma, C., Wang, F., Yao, G. and Jing, Z. Catalytic activity of Ni_3S_2 and effects of reactor wall in hydrogen production from water with hydrogen sulphide as a reducer under hydrothermal conditions. *Applied Energy* 2013; 104: 306–309.
47. Nakamura, R., Takashima, T., Kato, S., Takai, K., Yamamoto, M. and Hashimoto, K. Electrical current generation across a Black Smoker chimney. *Angewandte Chemie International Edition* 2010; 49: 7692–7694.
48. Yamaguchi, A., Yamamoto, M., Takai, K., Ishii, T., Hashimoto, K. and Nakamura, R. Electrochemical CO_2 reduction by Ni-containing iron sulfides: How is CO_2 electrochemically reduced at bisulfide-bearing deep-sea hydrothermal precipitates? *Electrochimica Acta* 2014; 141: 311–318.
49. Li, Y., Yamaguchi, A., Yamamoto, M., Takai, K. and Nakamura, R. Molybdenum sulfide: A bioinspired electrocatalyst for dissimilatory ammonia synthesis with geoelectrical current. *The Journal of Physical Chemistry C* 2017; 121: 2154–2164.
50. Zhang, Z. and Yates, J. T. Band bending in semiconductors: Chemical and physical consequences at surfaces and interfaces. *Chemical Reviews* 2012; 112: 5520–5551.
51. Shen, Z., Zhang, Y. L. and Jin, F. M. From $NaHCO_3$ into formate and from isopropanol into acetone: Hydrogen-transfer reduction of $NaHCO_3$ with isopropanol in high-temperature water. *Green Chemistry* 2011; 13: 820–823.

52. Shen, Z., Zhang, Y. L. and Jin, F. M. The alcohol-mediated reduction of CO_2 and $NaHCO_3$ into formate: A hydrogen transfer reduction of $NaHCO_3$ with glycerine under alkaline hydrothermal conditions. *RSC Adv* 2012; 2: 797–801.
53. Shen, Z., Gu, M., Zhang, M., Sang, W., Zhou, X., Zhang, Y. and Jin, F. The mechanism for production of abiogenic formate from CO_2 and lactate from glycerine: Uncatalyzed transfer hydrogenation of CO_2 with glycerine under alkaline hydrothermal conditions. *RSC Adv* 2014; 4: 15256–15263.
54. Su, J., Yang, L., Yang, X., Lu, M., Luo, B. and Lin, H. Simultaneously converting carbonate/bicarbonate and biomass to value-added carboxylic acid salts by aqueous-phase hydrogen transfer. *ACS Sustainable Chemistry & Engineering* 2015; 3: 195–203.
55. Kishida, H., Jin, F. M., Zhou, Z. Y., Moriya, T. and Enomoto, H. Conversion of glycerin into lactic acid by alkaline hydrothermal reaction. *Chemistry Letters* 2005; 34: 1560–1561.
56. Ramírez-López, C. A., Ochoa-Gómez, J. R., Gil-Río, S., Gómez-Jiménez-Aberasturi, O. and Torrecilla-Soria, J. Chemicals from biomass: Synthesis of lactic acid by alkaline hydrothermal conversion of sorbitol. *Journal of Chemical Technology & Biotechnology* 2011; 86: 867–874.
57. Costine, A., Loh, J. S. C., Busetti, F., Joll, C. A. and Heitz, A. Understanding hydrogen in Bayer process emissions. 3. Hydrogen production during the degradation of polyols in sodium hydroxide solutions. *Industrial & Engineering Chemistry Research* 2013; 52: 5572–5581.
58. Buhler, W., Dinjus, E., Ederer, H. J., Kruse, A. and Mas, C. Ionic reactions and pyrolysis of glycerol as competing reaction pathways in near- and supercritical water. *The Journal of Supercritical Fluids* 2002; 22: 37–53.
59. Ott, L., Bicker, M. and Vogel, H. Catalytic dehydration of glycerol in sub- and supercritical water: A new chemical process for acrolein production. *Green Chemistry* 2006; 8: 214–220.
60. Jin, F., Wang, Y., Zeng, X., Shen, Z. and Yao, G. Water under high temperature and pressure conditions and its applications to develop green technologies for biomass conversion. In *Application of Hydrothermal Reactions to Biomass Conversion*, Jin, F., Ed. Berlin, Heidelberg: Springer, 2014; pp. 3–28.

3 Hydrothermal Water Splitting for Hydrogen Production with Other Metals

Xu Zeng, Heng Zhong, Guodong Yao, and Fangming Jin

CONTENTS

3.1 Introduction 37
3.2 Hydrogen Production with Al...38
3.3 Hydrogen Production with Zn...40
3.4 Hydrogen Production with Mg..43
3.5 Hydrogen Production with Boron..43
3.6 Conclusions..44
Acknowledgment 44
References..44

3.1 INTRODUCTION

Extensive techniques have been developed to produce hydrogen from water by photochemical, electrochemical, or thermochemical routes. Although such processes are commercially available, hydrogen production with metals is still promising, owing to the reaction properties of in situ produced active hydrogen. One of the prospective methods of hydrogen generation is from water by means of metals or alloys. Besides iron, the appropriate materials for this goal include Al, Zn, Mg, B, and so on, which possess properties such as high efficiency, availability, and environmental safety. However, it is worthwhile to investigate the reactivity of water with metals for practical considerations, such as the cost of powders, the energy cycle economics, or the environmental impact. A systematic study is not available in the literature to date.

The aim of this review is to introduce the approaches and strategies of hydrogen production with metals (Al, Zn, Mg, and B) under mild hydrothermal conditions, which will be beneficial for promising methods of hydrogen generation from water in practical considerations.

3.2 HYDROGEN PRODUCTION WITH Al

The generation of hydrogen from aluminum–water reactions has gained growing interest, which can then be utilized in fuel cells, generators, and other devices [1]. Al is found to be the most appropriate for hydrogen production because of its highly positive ion (+3), low price, and abundance in the Earth's crust. Recycling of Al and aluminum oxide (i.e., into renewable electricity) is also advanced [2,3]; this is common in the aluminum industry [4]. Thus, Al is used as an ideal recyclable carrier of renewable energy [5]. To date, the exact mechanism of the Al–water reaction is not fully understood. Normally, it is considered that the reaction follows the scheme shown in Equation 3.1 [6]:

$$2Al + 6H_2O \rightarrow 2Al(OH)_3 + 3H_2 \tag{3.1}$$

Yavor et al. [7] studied the hydrogen production with the Al–water system for nano- and micron-sized spherical aluminum powders over the 20–200°C temperature range. The results showed that the maximum hydrogen production rate increases with the increase of temperature and the decrease of particle size, which is consistent with a surface reaction controlled by Arrhenius kinetics. The reaction is inhibited after it progresses a certain depth into the particles, that is, the penetration thickness. Figure 3.1 shows the hydrogen yield as a function of time. Higher reaction temperatures increased the reaction rate and shortened the induction time. The use of high-temperature Al–water reactions allows the use of micron-sized particles, rather than nano-powders.

Sharipov et al. studied the reaction of the Al atom in the ground electronic state with an H_2O molecule with quantum chemical calculations [8]. The reaction pathway is proposed as follows: in the first step, Equation 3.2 proceeded, leading to the aluminum hydrate formation in trans-isomeric form, which is twofold degenerate. Then, the HAlOH^{trans} structure can transform into the HAlOH^{cis} structure in the course of internal rotation (Equation 3.3). The HAlOH intermediate in both cis and trans

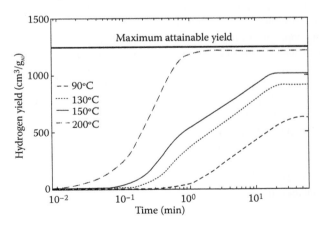

FIGURE 3.1 Hydrogen yield versus time at various temperatures.

configurations can dissociate into AlOH and H without passing through any saddle point (Equation 3.4). Thus, it was determined that the products AlO and H_2 form in the course of direct reaction path (Equation 3.5).

$$Al + H_2O \rightarrow TS_1 + HAlOH^{trans} \qquad (3.2)$$

$$HAlOH^{trans} \rightarrow TS_2 \rightarrow HAlOH^{cis} \qquad (3.3)$$

$$HAlOH^{cis/trans} \rightarrow AlOH + H \qquad (3.4)$$

$$AlOH + H \rightarrow TS_3 \rightarrow AlO + H_2 \qquad (3.5)$$

The zero point energy correction to the total energy of species was calculated. The normal frequency values predicted at the B3LYP/6-311+G(3df,2p) level of theory were applied. As shown in Figure 3.2, the energy excess of exothermal elementary process R1 must be released into vibrations of the HAlOH intermediate. It is worth noting that because of the small inversion barrier separating trans and cis isomers of HAlOH (~2 kJ/mol at the G3 level of theory) and the high energy of its vibrational excitation (~200 kJ/mol) after passing through TS1, there is no need to consider vibrationally excited $HAlOH^{trans}$ and $HAlOH^{cis}$ forms separately. Their calculations of activation barriers exhibited that the reaction $Al + H_2O \rightarrow AlOH + H$ occurs with smaller activation energy (E_a = 1830 K) than the reaction $AlOH + H \rightarrow AlO + H_2$ (E_a = 9450 K). The possibility of HAlOH stable molecule formation during direct interaction of an Al atom with an H_2O molecule was found to be negligible at moderate pressures. The reaction $Al + H2O \rightarrow AlOH + H$ proceeds much faster than the next step of the $Al + H_2O$ reaction path $AlOH + H \rightarrow AlO + H_2$.

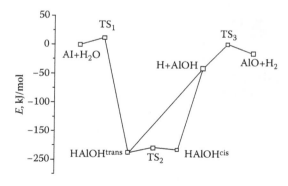

FIGURE 3.2 Relative energies with zero point correction for reactants, saddle points, intermediates and products on the PES of the $Al + H_2O$ system at the B3LYP/6-311+G(3df,2p) level of theory. All values are given with respect to the energy of the $Al + H_2O$ system.

3.3 HYDROGEN PRODUCTION WITH Zn

Recently, Zn has gained much attention in the hydrogen production from water under hydrothermal conditions, because the solar thermochemical cycle based on ZnO/Zn redox reactions achieved high energy conversion efficiency. Water splitting via the two-step ZnO/Zn thermochemical cycle has been shown as a promising and environmentally benign method for the production of hydrogen. As shown in Figure 3.3, Bhaduri studied the second step of the cycle, which is the exothermic hydrolysis reaction of Zn nanoparticles with steam [9]. The water-splitting reaction was performed at different steam and N_2 flow rates and reaction temperatures. The production rate and yield of H_2 at a reaction temperature of 600°C were calculated as 1.66×10^{-6} mol/g·s and 80%, respectively. The Zn-CNFs are potential for the H_2 production step of the ZnO/Zn thermochemical cycle.

Bhaduri discussed the effect of reaction temperature on the production of H_2 from steam. As shown in Figure 3.4, the amount of H_2 produced increases with increasing temperature between 400°C and 600°C. H_2 yield is ~42% at 400°C. The yield increases significantly to ~70%, as the reaction temperature is increased to 500°C. Thereafter, the increase in yield is small; for example, ~80% yield is obtained at 600°C. Their results illustrated that the method of in situ producing and dispersing Zn nanoparticles in the ACF/CNF support obviates these disadvantages and yields relatively larger H_2 yield and production rate in the hydrolysis step of the ZnO/Zn water splitting thermochemical cycle.

To date, there is scant literature about Zn–water reaction. Weidenkaff et al. [10] analyzed the thermogravimetric process of the hydrolysis of commercial zinc powder and solar zinc powder at temperatures of 350–500°C. Berman and Epstein [11] studied the kinetics of hydrogen production via the hydrolysis of liquid zinc through a bath at 450–520°C. Further investigations should be performed to understand the reaction process and improve the hydrogen production.

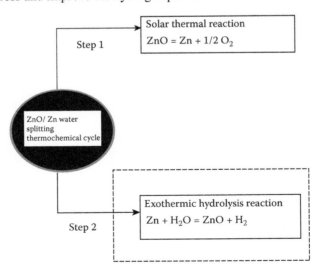

FIGURE 3.3 Steps involved in the ZnO/Zn water splitting thermochemical cycle. (From B. Bhaduri et al., *Chem. Eng. and Design*, 2014, 92, 1079–1090.)

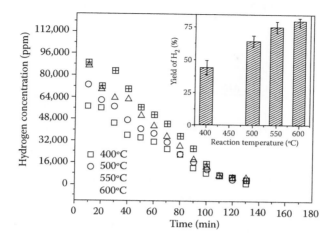

FIGURE 3.4 Effect of reaction temperature on the production of H_2 from steam.

Zeng et al. [12] reported a theoretical study with quantum chemical calculations based on experimental results to understand the highly efficient reduction of CO_2 to formic acid by using zinc under hydrothermal conditions. As shown in Figure 3.5, results showed that zinc hydride (Zn–H) is a key intermediate species in the reduction of CO_2 to formic acid, which demonstrates that the formation of formic acid is through an SN2-like mechanism. The reaction pathway could be understood as follows: in the first step, the Zn–$H^{\delta-}$ species is produced via the reaction of Zn and H_2O. In the second step, $H^{\delta-}$ attacks the carbon center $C^{\delta+}$, forming a bond with carbon as a transition state. In the final step, the OH^- species leave, forming $HCOO^-$ (an SN2-like mechanism). These results are useful to understand the metal–water reaction and the production of chemicals with in situ produced hydrogen.

Lv and Liu studied the thermodynamic process of hydrogen production via zinc hydrolysis [13]. The calculation was performed with the computational thermochemistry software FactSage. The ΔH^0 and ΔG^0 of zinc hydrolysis reaction under different temperatures at 1 atm pressure are shown in Figure 3.6. As shown in Figure 3.6, the

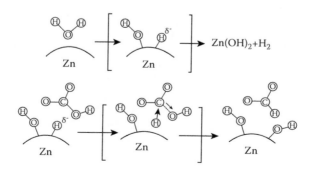

FIGURE 3.5 Proposed SN2-like mechanism for the formation of $HCOO^-$ with in situ produced hydrogen.

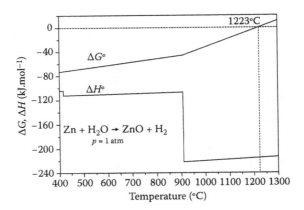

FIGURE 3.6 Temperature variations of ΔH^0 and ΔG^0 for the hydrolysis reaction of zinc at 1 atm.

hydrolysis of zinc is a moderate heat-releasing reaction. When the reaction temperature is lower than 850°C, the stoichiometric reactants can be heated to the reaction temperature by the heat revealed by the reaction itself in ideal state.

As shown in Figure 3.7, the influence of system pressure is much smaller. From the sight of reaction thermodynamics, it is reasonable to control the reaction temperature under 900°C to obtain a high hydrogen production. The result also shows that the increase of pressure can expand the high edge of zinc hydrolysis temperature window. The thermodynamic calculation result shows that the low edge of the water splitting temperature window is usually close to the high edge of the zinc hydrolysis temperature window. Under a pressure of 0.1 atm, the hydrogen equilibrium yield ratio decreases from 100% to 0% while the temperature increases from 500°C to 1500°C; the hydrogen production was from the hydrolysis of zinc.

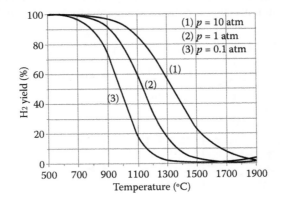

FIGURE 3.7 Variation of the H_2 equilibrium yield ratio as a function of the temperature and pressure.

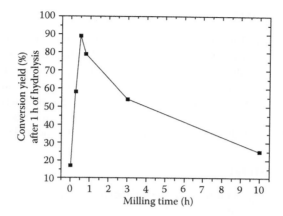

FIGURE 3.8 Conversion yield of Mg powders in 1 M KCl.

3.4 HYDROGEN PRODUCTION WITH Mg

It is very easy to produce hydrogen by using cheap Mg material with the following reaction:

$$Mg + 2H_2O \rightarrow Mg(OH)_2 + H_2 \qquad (3.6)$$

Grosjean et al. studied the hydrogen generation by hydrolysis of Mg and MgH$_2$ in pure water and 1 M KCl [14]. It was found that the hydrolysis reaction is faster and extensive. A significant increase of the hydrogen production is observed with the 30-min milled Mg sample. The possible reason is that the accentuation in the pitting corrosion resulted in the creation of numerous defects and fresh surfaces through the milling process.

When the hydrolysis of Mg is performed in 1 M KCl aqueous solution rather than pure water, the results are drastically different for Mg powders. Ball milling has a significant influence on Mg powder reactivity compared to hydrolysis in pure water. As shown in Figure 3.8, the 30-min milled Mg powder displays the best efficiency with a conversion yield reaching 89%. The conversion yield after 1 h of hydrolysis in 1 M KCl is plotted versus the Mg milling duration. The hydrolysis curves for milled Mg in KCl solution presents a distinctive shape with the appearance of an induction period.

3.5 HYDROGEN PRODUCTION WITH BORON

Hamed et al. [15] investigated the effect of temperature on boron hydrolysis (2B + 3H$_2$O → 3H$_2$ + B$_2$O$_3$) and studied the effect of boron oxide gasification reaction [B$_2$O$_3$(l) + 3H$_2$O(g) → 2H$_3$BO$_3$(g)] on the hydrolysis process. The hydrolysis experiments show parallel processes of boron hydrolysis and boron oxide gasification. X-ray analysis of the particles remaining in the crucible shows 100% boron oxide. The chemical reaction between the boron and the steam is much faster than the chemical reaction between the boron oxide and the steam.

3.6 CONCLUSIONS

Hydrogen is considered to be a promising alternative fuel and a raw material to produce different chemicals. One promising method of hydrogen generation is from water by means of metals or alloys. Besides iron, the appropriate material for this goal includes Al, Zn, Mg, B, and so on, which possesses properties such as high efficiency, availability, and environmental safety. The hydrogen production with metals (Al, Zn, Mg, and B) under mild hydrothermal conditions is introduced in this paper, which will benefit the hydrogen production from water in practical considerations.

ACKNOWLEDGMENT

The authors gratefully acknowledge the financial support of the National Natural Science Foundation of China (No. 21277091).

REFERENCES

1. H. Z. Wang, D. Y. Leung, M. K. Leung, M. Ni. A review on hydrogen production using aluminum and aluminum alloys. *Renew Sustain Energy Rev.*, 2009, 13, 845–853.
2. E. Balomenos, D. Paniasa, I. Paspaliaris, A. Steinfeld, E. Guglielmini, B. Friedrich. Carbo-thermic reduction of alumina: A review of developed processes and novel concepts. *Proceedings of EMC, Weimar, Germany*, 2011.
3. E. Balomenos, D. Paniasa, I. Paspaliaris. Energy and energy analysis of the primary aluminium production processes: A review on current and future sustainability. *Miner Process Extr Metall Rev.*, 2011, 32(2), 69–89.
4. F. Franzoni, M. Milani, L. Montorsi, V. Golovitchev. Combined hydrogen production and power generation from aluminum combustion with water: Analysis of the concept. *Int J Hydrogen Energy*, 2010, 35(4), 1548–1559.
5. D. Wen. Nanofuel as a potential secondary energy carrier. *Energy Environ. Sci.*, 2010, 3, 591–600.
6. J. Petrovic, G. Thomas. *Reaction of Aluminum with Water to Produce Hydrogen*. U.S. Department of Energy, 2008.
7. Y. Yavor, S. Goroshin, M. J. Bergthorson, D. L. Frost, R. Stowe, S. Ringuette. Enhanced hydrogen generation from aluminum–water reactions. *Int J Hydrogen Energy*, 2013, 38, 14992–15002.
8. A. Sharipov, N. Titova, A. Starik. Kinetics of Al + H_2O reaction: Theoretical study. *J. Phys. Chem. A*, 2011, 115, 4476–4481.
9. B. Bhaduri, N. Verma. A zinc nanoparticles-dispersed multi-scale web of carbon micro-nanofibers for hydrogen production step of ZnO/Zn water splitting thermochemical cycle. *Chem. Eng. and Design*, 2014, 92, 1079–1090.
10. A. Weidenkaff, A. W. Reller, A. Wokaun, A. Steinfeld. Thermogravimetric analysis of the ZnO/Zn water splitting cycle. *Thermochim Acta*, 2000, 359, 69–75.
11. A. Berman, M. Epstein. The kinetics of hydrogen production by oxidation of liquid zinc with water vapor. *Int J Hydrogen Energy*, 2000, 25, 957–967.
12. X. Zeng, M. Hatakeyama K. Ogata et al. New insights into highly efficient reduction of CO_2 to formic acid by using zinc under mild hydrothermal conditions: A joint experimental and theoretical study. *Phys. Chem. Chem. Phys.*, 2014, 16, 19836–19840.
13. M. Lv, H. Q. Liu. Thermodynamic analysis of hydrogen production via zinc hydrolysis process. *Com. Model. New Tech.*, 2014, 18, 273–277.

14. M. H. Grosjean, M. Zidoune, L. Roué, J.-Y. Huot. Hydrogen production via hydrolysis reaction from ball-milled Mg-based materials. *Int. J. Hydro. Energy*, 2006, 31, 109–119.
15. T. A. Hamed, B. Wahbeh, R. Kasher. The effect of a boron oxide layer on hydrogen production by boron hydrolysis. *World Renewable Energy Congress*, 2011, 1143–1149.

4 Hydrothermal Water Splitting for Hydrogen Production with Iron

Xu Zeng, Heng Zhong, Guodong Yao, and Fangming Jin

CONTENTS

4.1 Introduction 47
4.2 Hydrogen Production with Metals..48
4.3 Hydrogen Production with Iron 52
4.4 Hydrogen Production with Iron, Assisted by Carbonate Ions 55
4.5 Hydrogen Production by the Redox of Iron Oxide with Metal Additives 57
4.6 Conclusions...57
Acknowledgment 58
References...59

4.1 INTRODUCTION

Recently, the continuous rapid increase of CO_2 emissions has attracted much attention with respect to energy sustainability and the influence on the environment [1]. The Earth's surface temperature has risen by about 0.85°C from 1880 to 2012 according to the Intergovernmental Panel on Climate Change (IPCC 2013) [2]. The increase of CO_2 concentration is considered to harm the natural balance and threaten the ecological environment [3,4]. However, to meet the energy demands of economic growth and civilization, human activities continue to produce an excess amount of CO_2 to the carbon cycle. Therefore, the quest for a clean, carbon-free energy source to replace fossil fuels has become important.

One of the proposed solutions, which can be used to decrease the consumption of fossil fuels, is the use of hydrogen. Hydrogen can be used as a raw material to produce different chemicals, and it can be used as a key reagent in refineries [5,6]. More importantly, hydrogen is considered to be a promising alternative fuel. The hydrogen fuel cell-based car is used without emitting air pollutants such as CO_2, NO_x, and SO_x. Global hydrogen consumption in different industries can be seen in Figure 4.1. There are many pathways to produce hydrogen. The available commercial methods include the use of fossil fuels, water, or biomass as resources [7]. However, 98% of the annual globally produced hydrogen gas is obtained via the reforming of hydrocarbons due to the high effectiveness [8,9]. Alternative technologies for hydrogen

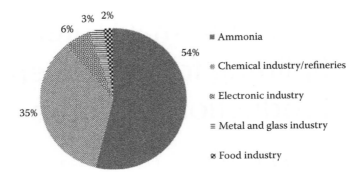

FIGURE 4.1 Global hydrogen consumption per industry. (From R. Chaubey, S. Sahu, O. O. James, S. Maity. *Renew Sust Energy Rev*, 2013, 23, 443–462.)

production have also been investigated extensively [10], such as biocatalysis, photocatalysis, and biophotocatalysis [11,12]. A new method of hydrogen production that is more clean and green is still strongly desired.

Thermochemical water decomposition has several advantages relative to other hydrogen production processes [13], for example, the sulfur–iodine thermochemical cycle [14], the addition of Zr in CeO_2 for the hydrogen reduction [15], and the potential of mixed conducting membranes to produce hydrogen by water dissociation [16]. In particular, high-temperature water has unique features compared to other processes. The ion product (K_w) at 250–300°C is approximately three orders of magnitude higher than that of ambient liquid water [17] and has been proven to be an environmentally friendly reaction media for the conversion of various types of biomasses [18,19]. One of the prospective methods of hydrogen generation is from water by means of metals or alloys. The appropriate material for this goal possesses properties such as high efficiency, availability, and environmental safety. However, it is worthwhile to investigate the reactivity of water with metals for practical considerations, such as the cost of the powders, the energy cycle economics, or the environmental impact. To date, a systematic study is not available in the literature.

The aim of this review is to introduce the approaches and strategies of hydrogen production with iron under mild hydrothermal conditions, which will benefit the promising methods of hydrogen generation from water in practical considerations.

4.2 HYDROGEN PRODUCTION WITH METALS

Jin et al. [20] reported that hydrogen production is demonstrated through metal oxidation in the presence of CO_2 with zero-valent Fe, Mn, Zn, and Al metals under hydrothermal conditions, where it is found that a maximum hydrogen formation yield of ca. 99.4% was achieved, as shown in Table 4.1. The metals, once oxidized, could be readily reduced (ca. 100% for Fe) to their zero-valent state. The results demonstrate that hydrogen production from water with metal could be used in a carbon cycle. The formation of H_2 from water under hydrothermal conditions in the presence of zero-valent metals Zn, Al, Fe, Mn, and Ni was examined. In the presence

TABLE 4.1
Hydrogen Formation in the Presence and Absence of CO_2[a]

Entry	Metal	NaHCO$_3$ [mmol]	Total[b]	CO$_2$[b]	H$_2$[b]	Y_{H_2}[d][%]
1	Ni	0	0	0	0	–
2	Ni	1	0	0	0	–
3	Fe	0	0	0	0	–
4	Fe[c]	0	0	0	0	–
5	Fe	1	50	2	48	28.1
6	Fe	2	80	2	75	44.4
7	Fe	3	100	4	90	53.8
8	Mn	0	97	1	90	–[e]
9	Mn	1	81	1	80	–[e]
10	Zn	0	52	0	52	–
11	Zn[c]	0	70	0	70	–
12	Zn	1	74	4	72	89.2
13	Zn	2	78	1	77	94.7
14	Al	0	58	0	58	–
15	Al[c]	0	114	0	114	–
16	Al	1	128	6	122	99.4

Gas [ml] header spans Total[b], CO$_2$[b], H$_2$[b].

[a] Reaction conditions: 573 K, 120 min. Fe, Ni = 6 mmol, Mn, Zn, Al = 4 mmol. pH of the initial solution was 8.3–9.0.
[b] Gas volumes corrected to 298 K and 1 atm.
[c] pH of the initial solution was adjusted to 9 using NaOH.
[d] Hydrogen yield, Y_{H_2} is defined as the ratio of the H$_2$ produced to that of stoichiometric H$_2$ produced assuming that Fe is oxidized into Fe$_3$O$_4$, Zn to ZnO, and Al to AlO(OH) (expressed as a percentage).
[e] For Mn, estimation of Y_{H_2} is uncertain since oxidation products include MnCO$_3$, Mn$_3$O$_4$ and MnO.

of CO$_2$, hydrogen was not produced when using Ni (Table 4.1, entry 2); however, hydrogen was formed in quantity when Fe was used. The results showed that in the metal–water reaction system, CO$_2$ can accelerate the production of hydrogen. For Mn, Zn, and Al, hydrogen was formed both in the absence and in the presence of CO$_2$ (Table 4.1). The CO$_2$ supplied in the form of bicarbonate readily promoted the formation of hydrogen from water with an approximately 99.4% maximum hydrogen formation yield.

Yavor et al. [21] introduced the reaction system with some commercially available metal powders and water at different temperatures. The authors investigated parameters such as the hydrogen production rate, yield, and reaction completeness. The authors found that hydrogen production was a function of temperature, although the results were influenced by the properties (e.g., average particle diameter and distribution, particle morphology, specific surface area, etc.) of the metal powders. The authors reported that the best metals to serve as energy carriers for hydrogen production are aluminum, magnesium, and manganese powders. For a metal, the theoretical amount of hydrogen produced from the metal–water reaction has no bearing, whether the reaction product is metal oxide or hydroxide, as demonstrated in Equations 4.1 and 4.2:

$$xM + yH_2O \rightarrow M_xO_y + yH_2 \qquad (4.1)$$

$$xM + 2yH_2O \rightarrow xM(OH)_{2y/x} + yH_2 \qquad (4.2)$$

The theoretical hydrogen yield from the metal–water reaction for various metals, shown in Figure 4.2a, provides a comparison of the different metals as potential candidates for hydrogen production. However, the maximum yield may not be reached. The energy released from the metal–water cycle, composed of both the metal–water reaction heat and the heat released from combustion of the hydrogen produced (shown in Figure 4.2b), can serve as another criterion for rating the different metals.

It can be seen that boron is the most attractive metal, gravimetrically as well as volumetrically. Other promising metals are aluminum, silicon, and magnesium, which are considered more reactive and show relatively high values of hydrogen production potential and heat release per unit mass. Heavier metals such as titanium, chromium, and manganese yield, theoretically, high volumetric hydrogen production and heat generated and may also be considered as potential energy carriers. Table 4.2 summarizes the data for all the experiments, conducted at four different

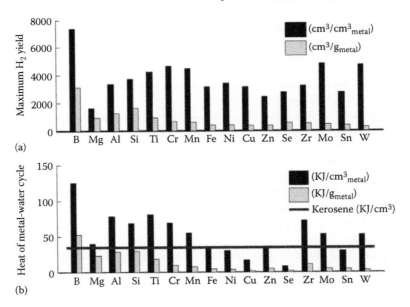

FIGURE 4.2 (a) Volumetric and gravimetric maximum theoretical hydrogen yield of the reaction of water with different metals. (b) Volumetric and gravimetric reaction heat of the metal–water cycle for the different metals.

TABLE 4.2
Total Yield and Measured Maximum Flow Rate with Different Metal Powders

	Total Yield [cm³/g]				Maximum Flow Rate [cm³/min/g]			
	80°C	120°C	150°C	200°C	80°C	120°C	150°C	200°C
B	97	195	453	560	25	21	203	217
Mg	171	295	344	921	24	140	287	2717
Al	277	362	717	1145	38	142	341	3043
Si	125	216	323	367	22	71	143	305
Ti	61	56	213	236	12	93	99	313
Cr	28	57	190	363	9	118	131	184
Mn	80	86	558	601	20	36	90	118
Fe	13	18	10	42	13	5	15	21
Ni	11	4	32	47	6	3	25	32
Cu	8	14	18	7	3	11	10	23
Zn	15	57	83	133	7	53	84	163
Se	21	0	29	32	6	1	32	50
Zr	18	36	97	252	5	97	93	127
Mo	5	26	40	56	3	3	28	83
Sn	35	82	118	150	6	68	97	95
W	3	36	37	61	2	21	24	52

temperatures with 16 different powders. This is then followed by a detailed analysis of the results.

In Figure 4.3, the hydrogen yield from experiments conducted at 120°C for all 16 powders illustrates that the powders with high specific energy, such as aluminum, magnesium, silicon, and boron, produce the largest amount of hydrogen per gram. Low-energy metal powders like copper, nickel, and selenium produce hardly any hydrogen and, in fact, fall within the experimental uncertainty limits for zero yields. The uncertainty limits are determined from the uncertainty in the flow meter readings, which is the largest source of uncertainty in the experiments.

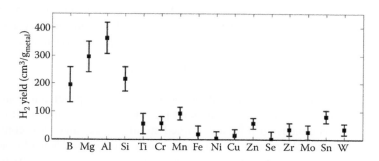

FIGURE 4.3 Hydrogen yield of experiments conducted at 120°C.

TABLE 4.3
Estimated Activation Energy for the Reaction of Different Metal Powders with Water

Metal Powder	E_a [kJ/mol]
Boron	30
Magnesium	53
Aluminum	50
Silicon	31
Titanium	36
Chromium	34
Manganese	22
Iron	8
Nickel	25
Copper	20
Zinc	35
Selenium	32
Zirconium	35
Molybdenum	42
Tin	31
Tungsten	38

Table 4.3 presents the estimated activation energy of the reaction of water with each of the 16 powders. Although only four data points are shown for each metal powder, it can be seen that they reasonably fit the Arrhenius equation, with an approximately linear behavior on the semilogarithmic graph. The slopes of the trend-line curves result in similar values of E_a for the aluminum and magnesium powders.

4.3 HYDROGEN PRODUCTION WITH IRON

The steam–iron process is a conventional technique for the generation of hydrogen gas by using the reaction of iron powder and water vapor [22]. As compared to the conventional steam–iron process, this process has the advantages of low temperature, simplicity, and high purity. Also, the direct generation of compressed hydrogen gas was favorable for its storage and utilization. The corresponding reaction equation is as follows:

$$3Fe + 4H_2O \rightarrow Fe_3O_4 + 4H_2 \tag{4.3}$$

Tsai et al. found that the compressed hydrogen gas has been generated from water and iron powders via a hydrothermal method [22]. The hydrothermal generation of hydrogen gas was conducted in a Teflon-lined stainless steel autoclave connected with a hydrogen gas collection bottle. In general, an appropriate amount of iron powders and 40 mL of pure water were put into the cylinder reactor (100 mL). By heating the reactor, hydrogen gas was generated. As shown in Figure 4.1, the vapor pressure of pure water remained at only about 1.5 bar. However, in the presence of iron powders, the pressure increased steadily with time and the pressure increase became more obvious with increasing the amount of iron powders, implying that hydrogen gas has been generated. By analyzing the composition of collected gas after reaction using gas chromatography, it was found that the water content was only about 0.525%. This revealed that the generated gas was hydrogen gas with high purity and demonstrated that the hydrothermal process developed in this work was indeed effective in the generation of compressed hydrogen gas with high purity. In addition, Figure 4.4 shows the x-ray powder diffraction (XRD) patterns of 3-mm iron powders before and after reaction at 120°C for 9 h. It was obvious that Fe has been partially converted into Fe_3O_4 after reaction, providing an evidence for the reaction.

To investigate the effects of size and morphology of iron powders on the generation of hydrogen gas, four kinds of iron powders were used. As shown in Figure 4.5, it was obvious that 100-nm iron powders exhibited a significantly faster generation rate than micro-sized iron powders. This could be attributed to the larger specific surface area of nano-sized iron powders that favored the reaction with water. However, it was noted that the 45-mm flat iron powders had a faster initial hydrogen generation rate than the 3-mm

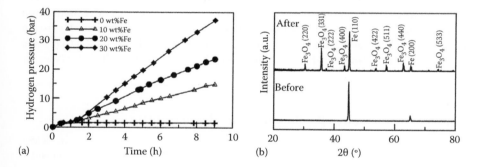

FIGURE 4.4 (a) Variation of pressure with time during the hydrothermal generation of hydrogen gas at 120°C and different amounts of 3-mm iron powder. (b) XRD patterns before and after reaction.

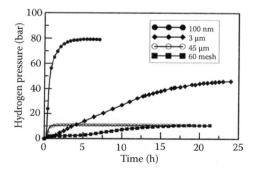

FIGURE 4.5 Variations of pressure with time during the hydrothermal generation of hydrogen gas in the presence of different iron powders (20 wt%) at 120°C.

spherical iron powders. This might be due to the fact that flat powders had a larger surface area than spherical powders. In addition, it was notable that the oxidation of iron powders occurs from the surface to the inner part. The resulting Fe_3O_4 shells might hinder the further oxidation of the inner part. Thus, the conversion might be affected by the reaction temperature, reaction time, and the size and morphology of iron powders.

Because of the advantages of low temperature, simplicity, high efficiency, high purity, and high pressure, such a novel and facile hydrogen gas generation technique might find potential applications in hydrogen energy-related devices and hydrogen gas-related chemical processes.

Wang et al. [23] reported a novel method for producing hydrogen from water with Fe as a reductant promoted by HS^- under mild hydrothermal conditions. Hydrogen production increased significantly in the presence of HS^-, compared to that in the absence of HS^-. The obvious hydrogen production was achieved in a low reaction temperature of 250°C and a very short reaction time (less than 2 h). The maximum yield of hydrogen production, which was defined as the percentage of produced hydrogen amount to theoretical one according to completive oxidation of Fe to Fe_3O_4 to produce hydrogen from water, was 34% at 300°C. HS^- may act as a catalyst and a possible HS^--catalyzed mechanism was proposed. This process may provide a promising solution for biomass-driven hydrogen production from water combined with the process of reducing iron oxide into its zero-valent state by bio-driven chemicals, such as glycerin. Hydrogen production was greatly enhanced to 6.7 mmol in the presence of HS^- for 2 h (see Table 4.4, entry 5), indicating that HS can improve hydrogen production. The analyses by

TABLE 4.4
Hydrogen Production in the Presence and Absence of HS^-

Entry	Temp. (°C)	Time (h)	Fe (mmol)	HS^- (mmol)	Conversion[a] (%)	H_2 Yield[b] (mmol)
1[c]	300	2	15	0	25.0	2.6
2	300	2	0	3	–	0.07
3	300	0[d]	15	3	50.3	5.4
4	300	1	15	3	54.8	5.9
5	300	2	15	3	62.0	6.7
6	300	4	15	3	60.2	6.8
7	300	6	15	3	57.0	6.7
8	250	1	15	3	23.5	2.7
9	300	2	15	6	74.9	10.2
10	300	2	15	1.5	–	0.39
11	300	2	15	0	–	0.01

[a] Conversion is defined as the percentage of amount of iron atom oxidized divided by the initial iron atom. These values were quantified by TOPAS software based on XRD patterns.
[b] Hydrogen yield is written as the average hydrogen production in amount (mmol).
[c] The pH of the solution was adjusted to 12.9.
[d] 0 h means the reactor was immediately took out for cooling after reaching the desired temperature without maintaining.

GC-TCD (gas chromatography with thermal conductivity detector) showed that there was only a peak assigned to hydrogen in the gas samples, indicating that the hydrogen produced was of high purity. Since the pH value of initial solution of HS^- was normally 12.7 because of hydrolysis of Na_2S, the hydrogen production may also be attributed to the effect of OH^- in the presence HS^-. To test this possibility, an experiment with Fe powder and 0.1 M NaOH (initial pH ~12.9) was conducted. As shown in Table 4.4, 2.6 mmol hydrogen was produced (entry 1) with a slight drop of pH value to 12.5 after the reaction, suggesting that OH^- has a promotion effect on Fe oxidation. However, the amount of hydrogen production of 2.6 mmol was much lower than that in the presence of HS (6.7 mmol). Thus, it is suggested that the increase in hydrogen production in the presence of HS^- is mainly attributed to the effect of HS^-.

4.4 HYDROGEN PRODUCTION WITH IRON, ASSISTED BY CARBONATE IONS

In Figure 4.6, Michiels's results confirmed that carbon dioxide can accelerate hydrogen production under hydrothermal conditions [24]. The hydrogen production takes place by reducing water through the oxidation of pure metallic iron powder Fe^0 to magnetite Fe_3O_4. The amount and purity of produced hydrogen gas were influenced by temperature, initial carbon dioxide pressure, and grain size of the metallic iron powder. The highest pure hydrogen (>99 mol%) with a percentage yield of approximately 80% is obtained by reacting zero-valent iron powder with a grain size of 5 μm at a temperature of 160°C and a carbon dioxide pressure of 6 bar (6 bar CO_2, 25°C, a volume of 60 mL: 15 mmol CO_2) for 16 h in a 100-mL autoclave that is filled for 40 mL with an aqueous potassium hydroxide solution of 1 M. The carbonate ion CO_3^{2-} has a key role in the hydrothermal system, acting as a catalyst, accelerating the hydrogen production. The oxidation of zero-valent iron to magnetite takes place via iron(II) carbonate $FeCO_3$ intermediates.

Figure 4.7 [25] suggests that the conversion of metallic iron in iron(II) carbonate becomes thermodynamically difficult, owing to a less negative difference in the Gibbs free energy. Because the Gibbs free energy has a function of temperature, the conversion of iron(II) carbonate in iron(II,III) oxide takes place preferably at

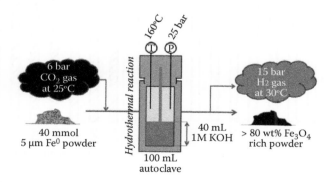

FIGURE 4.6 Michiels's scheme for the hydrogen production with iron, assisted by carbonate ions.

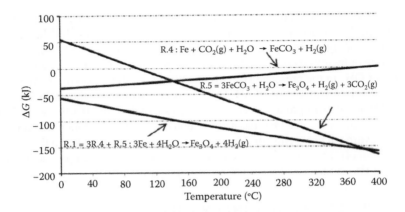

FIGURE 4.7 Change of standard Gibbs free energy as a function of temperature for the proposed occurring reactions. (From HSC Chemistry 5.0, Chemical Reaction and Equilibrium Software with Extensive Thermochemical Database. ver. 5.11, Outokumpu Research Oy, Pori, Finland, 2002.)

higher temperatures. Therefore, the optimum reaction temperature for the reaction of Fe^0 and H_2O can take place probably at lower temperatures, obtaining the highest percentage yield of hydrogen gas. The experiments with $p_i(CO_2) = 5$ bar, conducted in a temperature range from 140°C up to 260°C, with steps of 20°C, show that the optimal temperature is 160°C.

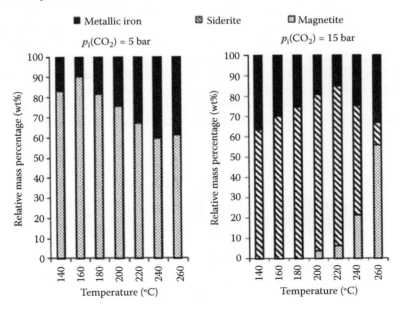

FIGURE 4.8 Mass percentages of the crystalline phases in the solid products obtained after hydrothermal reaction of metallic iron powder with a grain size of 100 μm at different temperatures.

In Michiels's study, XRD analyses demonstrate that up to three different crystalline phases are present in the solid product after reaction: metallic iron Fe^0, siderite $FeCO_3$, and magnetite Fe_3O_4. Semiquantitative analyses indicate that no amorphous phases are present. Figure 4.8 shows the different phases together with their relative mass percentages in the solid products that are obtained under different reaction conditions.

4.5 HYDROGEN PRODUCTION BY THE REDOX OF IRON OXIDE WITH METAL ADDITIVES

Liu and Wang investigated the redox performances of the modified Fe_2O_3 with single- or double-metal additives (Cr, Ni, Zr, Ag, and Mo; Mo–Cr, Mo–Ni, Mo–Zr, and Mo–Ag), which can be used for the hydrogen production by reduction of iron oxide with hydrogen and subsequent oxidation of reduced iron oxide with steam [26]. Mo-modified samples exhibited better redox performance than other samples, and the sample with Mo–Zr was the best for hydrogen production. The possible reason behind this is that the sintering of the particles was prevented because of the addition of Mo, and the decomposition of water was accelerated by the Mo cation before and after the redox reaction. The highest rate of hydrogen formation was from the fourth cycle at 300°C. The best cyclic stability of hydrogen storage/release is 360.9–461.1 µmol min^{-1} Fe-g^{-1} at operating temperature (no deactivation phenomenon after 10 cycles) and the high average capacity of hydrogen storage is 4.73% in four cycles. As shown in Table 4.5, the redox performances of the samples were notably elevated with the addition of a single metal. However, the effect of different metals on hydrogen production was different. The hydrogen production temperature of Fe_2O_3–Mo was much lower than that of the other samples. Therefore, the Mo additive had the most remarkable catalytic effect on improving hydrogen production.

Table 4.6 shows the cooperative effect of Mo with each of the other four metals (Zr, Cr, Ag, and Ni metal cations) as double-metal additives (Mo–Zr, Mo–Cr, Mo–Ag, and Mo–Ni) in the sample on the hydrogen production. As shwon in Table 4.6, the double-metal additives (Mo–M) in the samples could improve hydrogen production more remarkably than the corresponding single-metal additives (M) except the Mo additive. These double-metal additives in the samples can enhance hydrogen production significantly. According to the comparisons of all the Fe_2O_3 samples modified with single- and double-metal additives, the Mo additive shows good redox performances, and the Mo–Zr additive had the most remarkable role on improving the hydrogen production.

4.6 CONCLUSIONS

Hydrogen is considered to be a promising alternative fuel and a raw material in producing different chemicals. One promising method of hydrogen generation is from water by means of metals or alloys. The appropriate material for this goal includes

TABLE 4.5
The Redox Performances of Fe_2O_3 Modified with a Single Metal (Zr, Cr, Ag, Ni, and Mo)

| Sample | Cycle | Peak Temperature (°C) | H_2 Formation Temperature at 250 μmol min^{-1} Fe-g^{-1} | The Rate of H_2 Formation (μmol min^{-1} Fe-g^{-1}) ||| H_2(wt%) |
				At Peak Temperature (°C)	At 300 (°C)	At 354 (°C)	
Fe_2O_3-Zr	1st	510	441	796.8	78.4	165.2	4.45
	2nd	508	444	932.3	98.7	218.7	4.43
	3rd	495	419	822.9	130.4	234.9	4.75
	4th	510	370	640.4	102.1	216.2	4.52
Fe_2O_3-Cr	1st	453	394	518.8	159.1	285.2	4.33
	2nd	464	406	543.5	125.0	204.1	4.34
	3rd	466	376	627.1	126.6	242.6	4.45
	4th	488	410	616.4	137.5	203.1	4.55
Fe_2O_3-Ag	1st	517	432	963.6	41.6	99.7	4.76
	2nd	523	449	911.5	42.8	121.4	4.44
	3rd	544	455	758.3	61.5	126.1	4.70
	4th	506	435	659.8	61.9	114.1	4.68
Fe_2O_3-Ni	1st	336	411	354.1	100.0	277.8	4.51
	2nd	403	560	386.7	60.78	170.2	4.70
	3rd	385	522	312.3	87.5	207.8	4.67
	4th	497	511	353.4	77.1	135.4	3.63
Fe_2O_3-Mo	1st	333	270	670.8	487.5	616.6	4.74
	2nd	333	279	704.2	416.6	500.0	4.63
	3rd	333	279	654.9	435.2	385.5	4.64
	4th	323	277	453.1	453.1	395.8	4.78

Fe, Al, Zn, and so on, which possesses properties such as high efficiency, availability, and environmental safety. The approaches and strategies of hydrogen production with iron under mild hydrothermal conditions are mainly introduced in this paper, which will benefit the hydrogen production from water with metal in practical considerations.

ACKNOWLEDGMENT

The authors gratefully acknowledge the financial support of the National Natural Science Foundation of China (No. 21277091).

TABLE 4.6
The Redox Performances of Fe_2O_3 Modified with Double Metals Mo + M (Zr, Cr, Ag, and Ni)

Sample	Cycle	Peak Temperature (°C)	H_2 Formation Temperature at 250 μmol min⁻¹ Fe-g⁻¹	The Rate of H_2 Formation (μmol min⁻¹ Fe-g⁻¹)			H_2 (wt%)
				At Peak Temperature (°C)	At 300 (°C)	At 354 (°C)	
Fe_2O_3–Mo+Cr	1st	338	275	358.7	340.9	356.0	4.66
	2nd	340	277	441.7	331.6	312.5	4.62
	3rd	316	278	458.2	342.7	323.1	4.72
	4th	312	272	488.1	366.2	334.7	4.44
Fe_2O_3–Mo+Ag	1st	396	305	687.2	168.5	448.0	4.61
	2nd	332	285	756.8	350.3	523.0	4.65
	3rd	337	290	848.9	307.1	490.0	4.71
	4th	336	293	634.4	343.4	558.0	4.66
Fe_2O_3–Mo+Ni	1st	332	262	936.4	516.2	402.5	4.68
	2nd	291	274	565.2	562.5	427.1	4.53
	3rd	344	274	518.6	408.1	494.7	4.35
	4th	344	275	577.3	369.8	542.1	4.45
Fe_2O_3–Mo+Zr	1st	335	279	555.4	360.9	518.0	4.67
	2nd	338	276	875.3	396.1	439.0	4.74
	3rd	327	273	868.9	429.1	488.0	4.79
	4th	313	278	818.5	461.1	435.0	4.40
	5th	339	276	828.1	433.4	405.0	4.67
	6th	322	278	700.0	434.6	484.0	4.72
	7th	326	282	802.1	420.8	423.0	4.68
	8th	326	277	875.1	404.3	411.0	4.56
	9th	338	282	733.3	457.3	559.0	4.62
	10th	338	278	604.1	443.1	490.0	4.64

REFERENCES

1. A. C. Kone, T. Buke. Forecasting of CO_2 emissions from fuel combustion using trend analysis. *Renew Sustain Energy Rev*, 2010, 14, 2906–2915.
2. P. M. Cox, R. A. Betts, C. D. Jones, S. A. Spall, I. J. Totterdell. Acceleration of global warming due to carbon-cycle feedbacks in a coupled climate model. *Nature*, 2000, 408(6809), 184–187.
3. U. R. Sumaila, W. W. Cheung, V. W. Lam, D. Pauly, S. Herrick. Climate change impacts on the biophysics and economics of world fisheries. *Nat Clim Change*, 2011, 1(9), 449–456.
4. J. A. Patz, D. Campbell-Lendrum, T. Holloway, J. A. Foley. Impact of regional climate change on human health. *Nature*, 2005, 438(7066), 310–317.

5. M. Balat. Potential importance of hydrogen as a future solution to environmental and transportation problems. *Int J Hydrogen Energy*, 2008, 33, 4013–4029.
6. R. Chaubey, S. Sahu, O. O. James, S. Maity. A review on development of industrial processes and emerging techniques for production of hydrogen from renewable and sustainable sources. *Renew Sust Energy Rev*, 2013, 23, 443–462.
7. Y. Wang, F. Jin, X. Zeng, G. Yao, Z. Jing. A novel method for producing hydrogen from water with Fe enhanced by HS⁻ under mild hydrothermal conditions. *Int J Hydrogen Energy*, 2013, 38, 760–768.
8. O. Bicáková, P. Straka. Production of hydrogen from renewable resources and its effectiveness. *Int J Hydrogen Energy*, 2012, 37, 11563–11578.
9. G. Marbán, T. Valdés-Solís. Towards the hydrogen economy? *Int J Hydrogen Energy*, 2007, 32, 1625–1637.
10. B. Yildiz, M. S. Kazimi. Efficiency of hydrogen production systems using alternative energy technologies. *Int J Hydrogen Energy*, 2006, 31, 77–92.
11. H. Joo, Y. Jang, S. Lee, Y. G. Shu. Effects of constituents in photo/biocatalytic hydrogen production system using experimental design tool. *Ceram Trans*, 2006, 193(16), 139–154.
12. J. Yoon, H. Joo. Photobiocatalytic hydrogen production in a photoelectrochemical cell. *Korean J Chem Eng*, 2007, 24(5), 742–748.
13. M. A. Rosen. Advances in hydrogen production by thermochemical water decomposition: A review. *Energy*, 2010, 35, 1068–1076.
14. D. Doizi, V. Dauvois, J. L. Roujou, V. Delanne, P. Fauvet, B. Larousse. Total and partial pressure measurements for the sulphur–iodine thermochemical cycle. *Int J Hydrogen Energy*, 2007, 32(9), 1183–1191.
15. K. S. Kang, C. H. Kim, C. S. Park, J. W. Kim. Hydrogen reduction and subsequent water splitting of Zr-added CeO_2. *J Ind Eng Chem*, 2007, 13(4), 657–663.
16. U. Balachandran, T. H. Lee, S. Wang, S. E. Dorris. Use of mixed conducting membranes to produce hydrogen by water dissociation. *Int J Hydrogen Energy*, 2004, 29, 291–296.
17. R. W. Shaw, Y. B. Brill, A. A. Clifford, C. A. Eckert, E. U. Frank. Supercritical water—A medium for chemistry. *Chem Eng News*, 1991, 23, 26–39.
18. F. M. Jin, T. Moriya, H. Enomoto. Oxidation reaction of high molecular weight carboxylic acids in supercritical water. *Environ Sci Technol*, 2003, 37, 3220–3231.
19. A. Kruse, E. Dinjus. Hot compressed water as reaction medium and reactant: Properties and synthesis reactions. *J Supercrit Fluids*, 2007, 41, 361–379.
20. F. M. Jin, Y. Gao, Y. J. Jin. High-yield reduction of carbon dioxide into formic acid by zero-valent metal/metal oxide redox cycles. *Energy Environ Sci*, 2011, 4, 881–884.
21. Y. Yavor, S. Goroshin, J. M. Bergthorson, D. L. Frost. Comparative reactivity of industrial metal powders with water for hydrogen production. *Int J Hydrogen Energy*, 2015, 40, 1026–1036.
22. Y. C. Tsai, L. H. Liu, D. H. Chen. Hydrothermal generation of compressed hydrogen gas by iron powders. *RSC Adv*, 2016, 6, 8930–8934.
23. Y. Q. Wang, F. M. Jin, X. Zeng, G. D. Yao, Z. Z. Jing. A novel method for producing hydrogen from water with Fe enhanced by HSL under mild hydrothermal conditions. *Int J Hydrogen Energy*, 2013, 38, 760–768.
24. K. Michiels, J. Spooren, V. Meynen. Production of hydrogen gas from water by the oxidation of metallic iron under mild hydrothermal conditions, assisted by in situ formed carbonate ions. *Fuel*, 2015, 160, 205–216.
25. HSC Chemistry 5.0, Chemical Reaction and Equilibrium Software with Extensive Thermochemical Database. ver. 5.11, Outokumpu Research Oy, Pori, Finland, 2002.
26. X. J. Liu, H. Wang. Hydrogen production from water decomposition by redox of Fe_2O_3 modified with single- or double- metal additives. *Journal of Solid State Chemistry*, 2010, 183, 1075–1082.

5 Hydrothermal CO$_2$ Reduction with Iron to Produce Formic Acid

Jia Duo, Guodong Yao, Fangming Jin, and Heng Zhong

CONTENTS

5.1 Introduction 61
5.2 Experimental Section 63
    5.2.1 Materials 63
    5.2.2 Experimental Procedure 63
    5.2.3 Product Analysis 63
    5.2.4 Analytical Results of Liquid and Solid Samples 64
5.3 Results and Discussion 65
    5.3.1 Effect of Reaction Conditions on Fe Oxidation 65
    5.3.2 The Proposed Pathway for Fe Oxidation 68
    5.3.3 Effects of Reaction Conditions on the Formic Acid Yield 69
        5.3.3.1 Effects of the Amount of NaHCO$_3$ and Water Filling on the Formic Acid Yield 69
        5.3.3.2 Effects of Fe Amount and the Size of Fe Powder on the Formic Acid Yield 70
        5.3.3.3 Effects of Reaction Temperature and Time on the Formic Acid Yield 72
    5.3.4 The Proposed Mechanism of CO$_2$ Reduction with Fe to Produce Formic Acid 72
        5.3.4.1 The Role of the In Situ Hydrogen 72
        5.3.4.2 Catalytic Activity of the Formed Fe$_3$O$_4$ 73
        5.3.4.3 The Autocatalytic Mechanism 75
5.4 Conclusions 76
References 76

5.1 INTRODUCTION

CO$_2$, originating from fossil fuel combustion and other anthropogenic activities, has a strong impact on climate change and may pose catastrophic effects to humanity (Yang et al. 2011; Zhao et al. 2013); thus, the reduction of CO$_2$ emissions is an extensive, urgent, and long-term task. Rather than viewing CO$_2$ as a waste product, alternative approaches to convert CO_2 as a feedstock to more valuable chemicals show benefits. One approach to solving this problem is through artificial photosynthesis to recycle CO_2 directly into chemical energy by water splitting with solar energy (Hu 2012; Huber et al. 2003; Yadav et al. 2012). However, high conversion efficiencies of solar to fuels with the direct use of solar energy remains extremely challenging and it is far from reaching efficiencies close to application. In contrast to the direct use of solar energy, an integrated technology can be expected to have high potential for improving the efficiency of artificial photosynthetic systems. Some integrated technologies of solar hydrogen production and CO_2 reduction via a two-step water-splitting thermochemical cycle have been achieved, which use metal/metal oxide redox reactions such as Fe/Fe_3O_4 and Zn/ZnO (Chueh and Haile 2009; Galvez et al. 2008). Also, the electrochemical reduction of CO_2 can be regarded as a typical integrated technology to improve artificial photosynthetic efficiency based on the application to solar electricity.

In the Earth's crust and deep-sea hydrothermal vents, hydrothermal reactions have played an important role in the formation of fossil fuel and origin of life; for example, conversion of CO_2 dissolved into hydrocarbons in water, accompanying hydrothermal alteration of minerals (Horita and Berndt 1999; McCammon 2005). Thus, conversion of CO_2 to organics under hydrothermal conditions has always been an intense topic in the field of biochemistry and geochemistry at deep-sea vents (Lane and Martin 2012; Martin et al. 2008); however, there have been a very limited number of experiments that demonstrate the formation of organics. In proposed mechanisms of abiotic organic synthesis, the reaction of ferrous iron-bearing minerals with water is considered to generate reducing conditions (H_2) (McCollom and Seewald 2001). Hydrocarbon synthesis in geologic environments without solar radiation inspired us to use Fe as a reductant for hydrothermal CO_2 reduction under hydrothermal conditions because Fe is a cheap, abundant metallic and redox-active element on Earth; to explore an approach to efficiently reduce greenhouse CO_2 into organic chemicals; and to help understand the abiotic organic synthesis at deep-sea hydrothermal vents. Previous results showed that CO_2 could be converted into formic acid. More interestingly, we also found that a biomass-derived chemical of glycerin could achieve the reduction for the oxidation product Fe_3O_4 into Fe (Jin et al. 2011). Accordingly, an integrated technology to enhance the efficiency of artificial photosynthetic systems can also be expected by a redox of Fe/Fe_xO_y with biomass.

Formic acid is not only an important chemical in industry but also a useful reducing agent and source of carbon in synthetic chemical industries and has been traditionally employed as a preservative and an insecticide. Currently, the worldwide capacity for producing formic acid is about 800 kt/year (Moret et al. 2014). The demand for formic acid could continue to grow in the future, especially in the context of a renewable energy hydrogen carrier (Boddien et al. 2010; Uhm et al. 2008). At present, formic acid is produced from methanol and CO with a strong base. Therefore, formic acid production from greenhouse gases of CO_2 and water remains a topic of significant research.

Although previous study has showed potential of the reduction of CO_2 into formic acid with Fe in water, the yield of formic acid was only less than 2% without the

Hydrothermal CO_2 Reduction with Iron to Produce Formic Acid

addition of catalyst (Wu et al. 2009) and therefore further investigation for enhancing the efficiency of water splitting for CO_2 reduction is required.

The enhancement of Fe oxidation to produce a large amount of hydrogen should be focused on highly efficient CO_2 reduction into formic acid with Fe because many unreacted Fe were observed after reactions in previous research. Moreover, if large amounts of hydrogen can be produced, the surface of Fe_3O_4, a product of Fe oxidation, may be reduced in situ into Fe_3O_{4-x} and thus may result in the formed Fe_3O_4 showing a highly catalytic activity, leading to an autocatalytic process. Here, we report a simple and highly efficient strategy for water splitting to reduce CO_2 into formic acid by enhancing Fe oxidation with $NaHCO_3$ without the addition of any catalyst; the mechanism of this process is also discussed.

5.2 EXPERIMENTAL SECTION

5.2.1 Materials

Fe powder (325 mesh, ≥98% from Alfa Aesa; 100 mesh and 400 mesh, ≥98% Aladdin Chemical Reagent) was used in this study. $NaHCO_3$ (AR, ≥98% from Sinopharm Chemical Reagent Co., Ltd) was used as a CO_2 resource to simplify handling. Gaseous CO_2 and H_2 (>99.995%) were purchased from Shanghai Poly-Gas Technology Co., Ltd. In this study, deionized water was used in all experiments.

5.2.2 Experimental Procedure

The schematic drawing of the experimental setup has been described in detail elsewhere (Jin et al. 2001, 2005) and only a brief description is given below. The desired amount of $NaHCO_3$ (CO_2 source), reductant (Fe powder), and deionized water was loaded in a batch reactor. The reactor was made of a stainless steel 316 tubing (9.525 mm [3/8 in] o.d., 1 mm wall thickness and 120 mm length) with fittings sealed at each end, providing an inner volume of 5.7 mL. After loading, the reactor was immersed in a salt bath. During the reaction, the reactor was shaken and kept horizontally to enhance the mixture and heat transfer. After the preset reaction time, the reactor was removed from the salt bath to quench in a cold water bath. After the reactions, the liquid, gaseous, and solid samples were collected for analysis after cooling to room temperature. Water filling was defined as the ratio of the volume of the water put into the reactor to the inner volume of the reactor, and the reaction time was defined as the duration of time that the reactor was kept in the salt bath.

5.2.3 Product Analysis

Liquid samples were filtered (0.22 μm filter film) and then analyzed by high-performance liquid chromatography (HPLC, Aglient 1260), total organic carbon (TOC, Shimadzu TOC 5000A), and gas chromatography/mass spectroscopy (GC/MS, Agilent 7890). The solid samples were washed with deionized water three times to remove impurities and with ethanol three times to make the solid sample dry quickly. The samples were then dried in an isothermal oven at 40°C for 3–5 h and characterized using x-ray diffraction (XRD, Shimadzu XRD-6100), scanning electron microscopy (SEM, FEI Quanta 200), x-ray photoelectron spectroscopy (XPS, EXCALAB 250), and dispersive Raman microscopy (RAM, Senterra R200-L). The yield of formic acid was defined as the percentage of formic acid and the initial $NaHCO_3$ on a carbon basis. The selectivity of formic acid was defined as the ratio between the amount of carbon in formic acid and the total organic carbon in the samples by TOC analysis. The conversion of Fe is defined as the percentage of the amount of oxidized iron divided by the initial iron atom, which was quantified by MDI jade software based on XRD patterns.

5.2.4 Analytical Results of Liquid and Solid Samples

Liquid samples were analyzed by HPLC and GC/MS, as shown in Figure 5.1, and the main product was formic acid. Also, analytic results from GC/MS indicated that a small amount of acetic acid was detected. The selectivity for the production of formic acid was approximately 99% by TOC analysis. These results indicated that CO_2 could be reduced into formic acid with a high efficiency selectivity by only using Fe without any catalyst addition. Thus, we mainly focused on the effect of the formic acid yield in our study. To study the oxidation level of Fe, solid samples after reactions were analyzed by XRD. As shown in Figure 5.2, the detected peaks were mainly Fe_3O_4 and Fe.

Moreover, considering that Fe_3O_4 and γ-Fe_2O_3 have the same inverse spinel structure and similarity in their d spacing, to further confirm that the oxidation product is Fe_3O_4, rather than Fe_2O_3, further identification of Fe_3O_4 from Fe oxidization was conducted by Raman spectroscopy. As shown in Figure 5.3, the features of Fe_3O_4 appeared at 219, 284, 399, 490, and 604 cm^{-1}, which correspond to the stretching vibration mode of Fe–O in the crystal Fe_3O_4 (Oblonsky and Devine 1995).

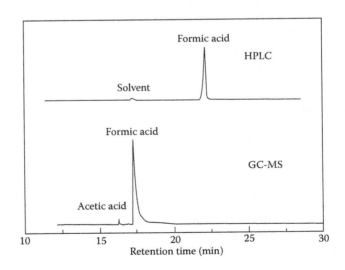

FIGURE 5.1 HPLC and GC-MS chromatogram of liquid sample after reaction (300°C, 2 h, 6 mmol $NaHCO_3$, 12 mmol Fe, 55% water filling).

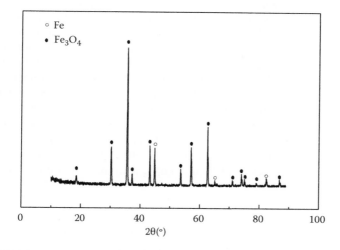

FIGURE 5.2 XRD pattern of solid residue after the reaction (300°C, 2 h, 6 mmol NaHCO$_3$, 12 mmol Fe, 55% water filling).

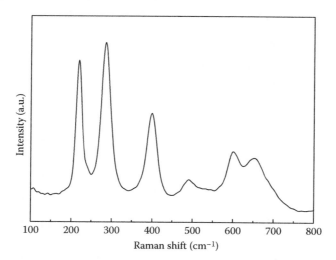

FIGURE 5.3 Raman scattering spectra of the oxidation product of Fe.

In addition, there are no typical γ-Fe$_2$O$_3$ peaks (~700 cm^{-1}), indicating the absence of γ-Fe$_2$O$_3$ (Shebanova and Lazor 2003).

5.3 RESULTS AND DISCUSSION

5.3.1 Effect of Reaction Conditions on Fe Oxidation

First, to enhance Fe oxidation for the hydrogen production from water, a series of experiments were conducted by changing a wide range of reaction parameters including reaction time, temperature, water filling, the size of iron powder, and the addition of NaOH and $NaHCO_3$. Results revealed that $NaHCO_3$ displayed a better performance for producing hydrogen (Table 5.1). These observations prompted us to further study the improving role of $NaHCO_3$ in hydrogen production by examining Fe oxidation in water with changing the amount of $NaHCO_3$. According to the XRD pattern of the solid residue after the reaction in Figure 5.4, Fe was rarely oxidized without $NaHCO_3$, but there was Fe oxidation in the presence of $NaHCO_3$. Thus, we conducted a series of experiments in which we varied the amount of $NaHCO_3$ from 1 to 6 mmol; the result showed that the peak of Fe became much smaller and the peak of Fe_3O_4 became stronger with the increase of $NaHCO_3$. Figure 5.5 shows a dramatic enhancement in the conversion of Fe into Fe_3O_4 with the increase in $NaHCO_3$, and

TABLE 5.1
Conversion of Fe with the Different Reaction Parameters

Entry	Reaction Parameters	State 1	Fe Conversion (%)	State 2	Fe Conversion (%)
1	Time (min)[a]	5	73.7	120	89.4
2	Temperature (°C)[b]	275	79	325	90.6
3	NaOH (mmol)[c]	0	46.5	1	47.7
4	$NaHCO_3$ (mmol)[d]	1	46.5	6	90.2

Note: Reaction conditions: 12 mmol Fe, 55% water filling.
[a] 300°C, 6 mmol $NaHCO_3$.
[b] 2 h, 6 mmol $NaHCO_3$.
[c] 300°C, 2 h, 1 mmol $NaHCO_3$.
[d] 300°C, 2 h.

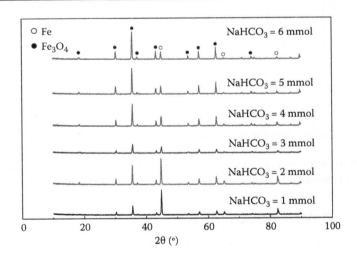

FIGURE 5.4 XRD pattern of solid residue after reaction with different amounts of $NaHCO_3$ (300°C, 2 h, 12 mmol Fe, 55% water filling).

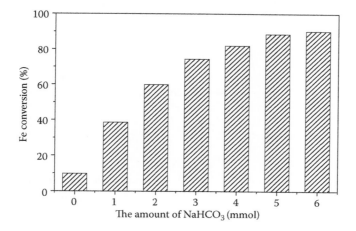

FIGURE 5.5 The conversion of Fe into Fe_3O_4 with different $NaHCO_3$ amounts (300°C, 2 h, 12 mmol Fe).

the conversion of Fe reached 90% with 6 mmol $NaHCO_3$. The conversion of Fe is defined as the percentage of the amount of oxidized iron divided by the initial iron atom, which was quantified by MDI jade software based on XRD patterns.

Then, the morphology of the solid residue was examined by SEM to further verify the promotion of $NaHCO_3$ on Fe oxidation. SEM imaging without $NaHCO_3$ exhibited many disorderly inclusions and only very few small irregular polyhedral particles on the surface of inclusions (Figure 5.6a). In combination with the results shown in Figure 5.4, the inclusions would be the unreacted Fe, and the small particles on the surface of inclusions would be the formed Fe_3O_4. In the presence of 1 mmol $NaHCO_3$, small particles of Fe_3O_4 became irregular polyhedral particles, while inclusions (the unreacted Fe) decreased (Figure 5.6b). When increasing $NaHCO_3$ to 6 mmol, Fe_3O_4, with irregular polyhedral particles, became an angular octahedral structure (Figure 5.6c). Clearly, the increase in $NaHCO_3$ led to the increase in the amount Fe_3O_4 and also improved the growth of the Fe_3O_4 crystal to a fine-grained crystal.

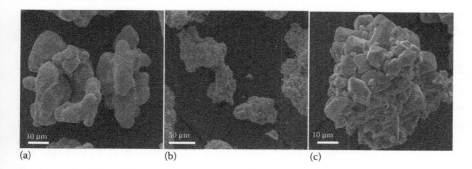

FIGURE 5.6 SEM images of the interface of solid residues with (a) 0 mmol $NaHCO_3$, (b) 1 mmol $NaHCO_3$, and (c) 6 mmol $NaHCO_3$ (300°C, 2 h, 12 mmol Fe, 55% water filling).

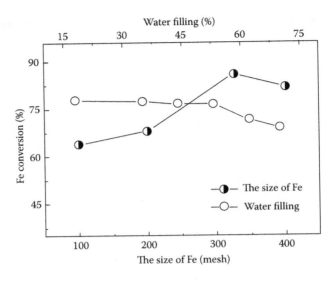

FIGURE 5.7 Effect of the size of Fe and water filling on Fe conversion (300°C, 2 h, 12 mmol Fe; 6 mmol NaHCO$_3$ and 55% water filling for the size of Fe; 1 M NaHCO$_3$ for water filling).

Subsequently, the effect of the size of Fe on Fe oxidation was further investigated. As shown in Figure 5.7, Fe conversion increased first and then decreased with the decrease in the size of Fe, and the highest Fe conversion 86.3% was obtained with 325-mesh Fe powder. Because a smaller Fe powder has a larger specific area on which to supply active surface for Fe oxidation, the small size of Fe will benefit Fe oxidation. However, Fe conversion showed a slight decrease at 400-mesh Fe powder; this is probably attributed to the active surface that started to saturate at 400 mesh and then slowed down Fe oxidation.

The increase of water filling can contribute to the increase in the pressure of this system. Hence, the experiment was conducted by changing water filling with a constant concentration of NaHCO$_3$ (1 M) to investigate the effect of the reactor pressure on Fe oxidation. The conversion of Fe into Fe$_3$O$_4$ initially remained unchanged and then dropped slightly (Figure 5.7). It is indicated that the effect of water filling is not obvious on Fe oxidation.

5.3.2 The Proposed Pathway for Fe Oxidation

The change in oxidation products of Fe with the reaction time was examined at different temperatures to understand the role of NaHCO$_3$ in improving Fe oxidation. At higher temperatures of 300°C and 325°C, only Fe$_3$O$_4$ was detected as an oxidative product of Fe, and as shown in Figure 5.8, the conversion of Fe into Fe$_3$O$_4$ proceeded very quickly in the first 5 min, then became slower from 5 to 30 min, and remained nearly constant after 90 min. However, at a lower temperature of 275°C, in addition to Fe$_3$O$_4$, a large number of FeCO$_3$ were detected in the first 5 min and the amount of FeCO$_3$ decreased gradually with the increase in reaction time, especially at

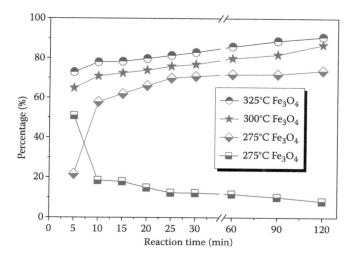

FIGURE 5.8 Effects of reaction time and reaction temperature on Fe oxidation (6 mmol NaHCO$_3$, 12 mmol Fe).

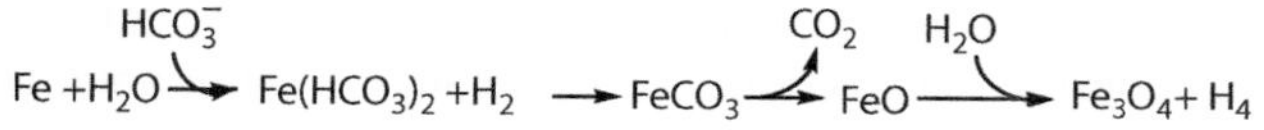

$$Fe + H_2O \xrightarrow{HCO_3^-} Fe(HCO_3)_2 + H_2 \longrightarrow FeCO_3 \xrightarrow{CO_2} FeO \xrightarrow{H_2O} Fe_3O_4 + H_4$$

FIGURE 5.9 The possible pathway of Fe oxidation in the presence of HCO$_3^-$.

5 to 10 min, accompanied by the rapid increase of Fe$_3$O$_4$. These observations suggest that Fe is oxidized into Fe$_3$O$_4$ via FeCO$_3$. It has been reported that FeCO$_3$ can be decomposed into FeO and CO$_2$ in both air and nitrogen atmospheres at relative high reaction temperatures (above 300°C) (Webb et al. 1970). According to these results, the mechanism of promoting Fe oxidation with NaHCO$_3$ can be explained in Figure 5.9. In the presence of HCO$_3^-$, Fe, NaHCO$_3$, and water first react to produce hydrogen with the formation of Fe(HCO$_3$)$_2$, which is then easily decomposed into FeCO$_3$ along with CO$_2$ and H$_2$O owing to the instability of Fe(HCO$_3$)$_2$ (Ogundele and White 1986). Subsequently, FeCO$_3$ is further transformed into FeO and CO$_2$. Since FeO is extremely unstable in water, FeO is further oxidized into Fe$_3$O$_4$ in water with the production of hydrogen.

5.3.3 Effects of Reaction Conditions on the Formic Acid Yield

5.3.3.1 Effects of the Amount of NaHCO$_3$ and Water Filling on the Formic Acid Yield

After understanding the role of HCO$_3^-$ in promoting Fe oxidization, the in situ CO$_2$ reduction into formic acid with Fe was investigated. Similarly, as shown in Figure 5.10, the initial amount of NaHCO$_3$ had a great impact on the yield of formic acid and the yield rapidly increased from 32% to 68.3% with the increase in NaHCO$_3$

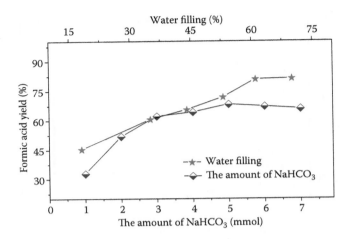

FIGURE 5.10 Effect of NaHCO$_3$ amount and water filling on the yield of formic acid (300°C, 2 h, 12 mmol Fe for NaHCO$_3$ concentration; 1 M NaHCO$_3$ for water filling).

from 1 to 5 mmol. An examination of pH in the solution indicated that the pH of the solution before and after the reactions was 8.3 and about 8.6, respectively, indicating that HCO$_3^-$ existed during the reaction. The moderately alkaline solution caused by the buffer capacity of HCO$_3^-$ easily leads to both the oxidization of Fe for producing hydrogen and the formation of formic acid. Thus, HCO$_3^-$ is crucial not only for the substrate but also for the pH buffer, allowing a longer reaction.

In the investigation on the effect of water filling/reaction pressure on Fe oxidation/hydrogen production in water, no significant change was observed (Figure 5.10), which is most likely because a higher pressure is not beneficial for the formation of a gaseous product of hydrogen (3Fe + 4H$_2$O → Fe$_3$O$_4$ + 4H$_2$). However, no gas product is produced during CO$_2$ hydrogenation (2H + 2HCO$_3^-$ → 2HCOO$^-$ + H$_2$O), and thus a higher pressure could be expected to benefit CO$_2$ hydrogenation. As expected, the yield of formic acid increased significantly with the increase in water filling, and the highest yield of about 81% was achieved with 60% water filling (Figure 5.10).

5.3.3.2 Effects of Fe Amount and the Size of Fe Powder on the Formic Acid Yield

The effect of Fe amount on CO$_2$ reduction was investigated under the optimized reaction conditions at hand. As shown in Figure 5.11, the formic acid yield was markedly improved with increasing Fe amount. As shown in Figure 5.11, a considerable formic acid yield can be achieved when using 16 mmol Fe with a lower Fe/NaHCO$_3$ ratio of 8:3. Considering the pressure limitation of the reactor used in this study, experiments with a higher Fe amount were not conducted. To further examine the effect of Fe amount, experiments with a lower amount of 2 mmol NaHCO$_3$ were conducted, and the yield of formic acid reached up to 92%.

The size of Fe powder is an important factor for Fe oxidation; thus, the effect of the size of Fe powder on the formic acid yield was subsequently investigated. We further

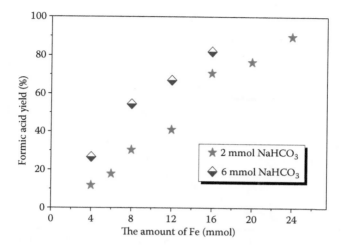

FIGURE 5.11 Effect of the amount of Fe on the yield of formic acid (300°C, 2 h).

performed experiments with the size of Fe ranging from 100 to 400 mesh, and the yield increased significantly from 100 to 325 mesh and then decreased slightly at 400 mesh (Figure 5.12), which was identical with the effect on Fe oxidation. The smaller Fe powder had a larger specific area on which to supply active surface for this reaction. Because the drop in the formic yield was less than 5% at an Fe size of 400 mesh, according to our results, the active surface probably started to saturate at 325 to 400 mesh and the reaction became slower. Therefore, we chose 325-mesh Fe powder in succeeding studies.

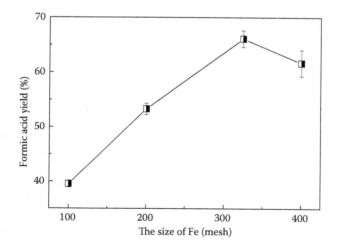

FIGURE 5.12 Effect of the size of Fe on the yield of formic acid (300°C, 2 h, 12 mmol Fe, 55% water filling).

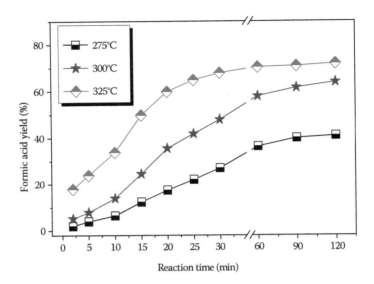

FIGURE 5.13 Effects of reaction time and reaction temperature on formic acid yield (6 mmol NaHCO$_3$, 12 mmol Fe).

5.3.3.3 Effects of Reaction Temperature and Time on the Formic Acid Yield

A kinetic curve of the formic acid production is shown in Figure 5.13. At first, the yield of formic acid increased slowly and then increased rapidly with the increase in reaction time. Finally, no significant change in formic acid was observed. Furthermore, CO$_2$ reduction proceeded efficiently at 300°C and 325°C, and a high formic acid yield of 40–60% was achieved in the first 30 min, indicating that a higher temperature is favorable for the production of formic acid. This observation is in agreement with results for the hydrothermal conversion of biomass, which may be related to the maximal ion-product constant of subcritical water occurring between 280°C and 300°C (Franck 1983; Jin et al. 2011).

5.3.4 THE PROPOSED MECHANISM OF CO$_2$ REDUCTION WITH Fe TO PRODUCE FORMIC ACID

5.3.4.1 The Role of the In Situ Hydrogen

Generally, CO$_2$ hydrogenation needs a catalyst for activating H$_2$ or CO$_2$. However, a high formic acid yield was obtained without adding any other catalyst in the present study, which suggests that some intermediates or products formed in situ such as FeCO$_3$ or Fe$_3$O$_4$ may act as a catalyst when using Fe as a reductant. To test this assumption, reactions with gaseous hydrogen (H$_2$) as a substitute for Fe were conducted in the presence of FeCO$_3$ or Fe$_3$O$_4$. As shown in Table 5.2, no significant change in the yield of formic acid with commercial Fe$_3$O$_4$ or FeCO$_3$ (entries 3 and 9) was observed compared to those without FeCO$_3$ or Fe$_3$O$_4$ (entry 2). Interestingly, however, the formic acid yield with the collected Fe$_3$O$_4$ after the reaction was higher (entry 4) than that with commercial Fe$_3$O$_4$ or FeCO$_3$, particularly for wet Fe$_3$O$_4$

TABLE 5.2
Effect of Additives of Fe_3O_4 or $FeCO_3$ on the Formic Acid Yield (300°C, 2 h)

Entry	NaHCO$_3$ (mmol)	H$_2$ (MPa)	Additives	Yield (%)
1	6	2		2.3
2	6	6		2.6
3	6	6	Commercial Fe$_3$O$_4$	0.6
4	6	6	Collected dry Fe$_3$O$_4$	4.5
5	6	6	Collected wet Fe$_3$O$_4$	14.3
6	6	–		0
7	–	–	Commercial FeCO$_3$	0
8	6	–	Commercial FeCO$_3$	1.7
9	6	6	Commercial FeCO$_3$	2.9

collected after the reaction (entry 5), indicating the catalytic activity of the Fe$_3$O$_4$ collected after the reaction in the reduction of CO$_2$ into formic acid. Although the formic acid yield with gaseous hydrogen in the presence of collected wet Fe$_3$O$_4$ increased, the yield was lower than that with Fe. These results suggest that the activity of the hydrogen produced in situ at high-temperature water with Fe may be higher than that of gaseous hydrogen in the reduction of CO$_2$. The study following this line of inquiry is now in progress.

5.3.4.2 Catalytic Activity of the Formed Fe$_3$O$_4$

Subsequently, to investigate the catalytic mechanism of collected Fe$_3$O$_4$, we analyzed chemical composition and valence states on the collected and commercial Fe$_3$O$_4$ by XPS. As shown in Figure 5.14, both collected and commercial Fe$_3$O$_4$ have two peaks at 710.5 and 724.6 eV corresponding to the binding energies of Fe 2p1/2 and Fe 2p3/2, respectively (Anderson et al. 1996; Yamashita and Hayes 2008). Generally, Fe$_3$O$_4$ has no satellite peak of Fe 2p$_{3/2}$ (Hawn and DeKoven 1987; Muhler et al. 1992), but commercial Fe$_3$O$_4$ has a weak satellite peak of Fe 2p$_{3/2}$ for Fe^{3+} at 718.7 eV as shown in Figure 5.14a (Yamashita and Hayes 2008), indicating the presence of more Fe^{3+} in the surface of commercial Fe$_3$O$_4$. It is most likely because some Fe^{2+} were oxidized into Fe^{3+} in the air on the surface of commercial Fe$_3$O$_4$. As shown in Table 5.3, which lists the Fe^{2+}/Fe^{3+} ratio of collected and commercial Fe$_3$O$_4$, the Fe^{2+}/Fe^{3+} ratio of commercial Fe$_3$O$_4$ was 0.22:0.78, which was higher than the theoretical ratio of Fe$_3$O$_4$ (0.33:0.67), further demonstrating the oxidation of Fe^{2+} into Fe^{3+}. Conversely, the Fe^{2+}/Fe^{3+} ratio of the collected Fe$_3$O$_4$ was 0.89:0.11, which was apparently lower than the theoretical ratio of Fe$_3$O$_4$. This discrepancy appears likely to be that Fe^{3+} on the Fe$_3$O$_4$ surface was in situ reduced into Fe^{2+} by the hydrogen formed from Fe oxidation owing to the enhancement in Fe oxidation with NaHCO$_3$.

The O1s spectrum of collected and commercial Fe$_3$O$_4$ is displayed at different BEs, Oa (529.7 eV), Ob (531.3 eV), and Oc (533.2 eV) in Figure 5.14d. The Oa, Ob, and Oc peaks are usually attributed to O^{2-} in the Fe$_3$O$_4$ structure and to oxygen of

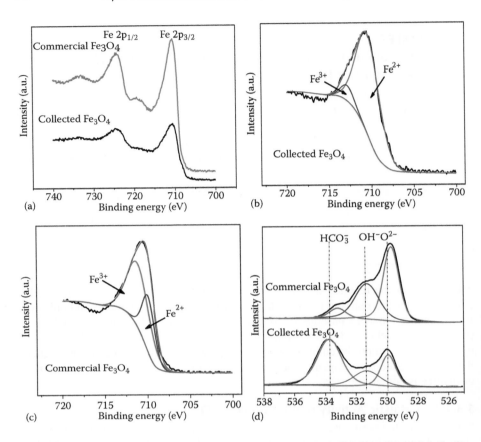

FIGURE 5.14 X-ray photoelectron spectroscopies of collected and commercial Fe_3O_4: (a) Fe 2p, (b) collected Fe_3O_4 $2p_{3/2}$, (c) commercial Fe_3O_4 $2p_{3/2}$, and (d) O1s.

TABLE 5.3
Ratio of Fe^{2+} and Fe^{3+} for Collected and Commercial Fe_3O_4

	Peak Position (eV)			Atomic Ratio (%)	
	Fe $2p_{1/2}$	Satellite	Fe $2p_{3/2}$	Fe^{2+}	Fe^{3+}
Collected Fe_3O_4	724.1	–	710.6	89	11
Commercial Fe_3O_4	724.1	718.6	710.6	22	78

OH groups on the Fe_3O_4 surface and are shown in Figure 5.14d. The O1s spectra can be fit into three peaks: chemisorbed or dissociated oxygen on the surface of the Fe_3O_4, such as $-CO_3/-HCO_3$, adsorbed H_2O, or adsorbed O_2 (Anderson et al. 1996; Maurice et al. 1996). Table 5.4 lists the relative intensity of Oa, Ob, and Oc peaks of the O1s spectrum for collected and commercial Fe_3O_4. Compared to commercial Fe_3O_4, the relative intensity of the Oa peak for collected Fe_3O_4 decreased, indicating

TABLE 5.4
Relative Intensity of Oa, Ob, and Oc Peaks of the O1s Spectrum of Collected and Commercial Fe_3O_4

	Peak Binding Energy (eV)			Relative Intensity (%)		
	Oa	Ob	Oc	Oa	Ob	Oc
Collected Fe_3O_4	529.9	531.4	533.8	33.3	19.7	47.0
Commercial Fe_3O_4	529.7	531.3	533.2	54.7	30.6	14.8

the formation of more oxygen vacancies. Meanwhile, the relative intensity of the Oc peak for collected Fe_3O_4 was stronger than that for commercial Fe_3O_4, suggesting the adsorption of more HCO_3^- on the surface of collected Fe_3O_4. More Fe^{2+} on the Fe_3O_4 surface should provide active sites for CO_2 reduction.

5.3.4.3 The Autocatalytic Mechanism

Combining XPS analysis with the kinetic curve (Figures 5.13 and 5.14), water splitting for CO_2 reduction with Fe in water should be a typical self-catalytic reaction, which includes a short induction and fast formation period. In the short induction period during the initial 5–10 min, hydrogen and Fe_3O_4 with more oxygen vacancies (Fe_3O_{4-x}) are formed. Then, the formed Fe_3O_{4-x} acts as a catalyst in the reduction of HCO_3^- into formic acid and thus leads to a quick increase in formic acid yield within 10–30 min. Finally, the reaction comes to an equilibrium stage. Thus, the profile of formic acid formation shows a sigmoidal-like curve. With these experimental results, a tentative mechanism of dissociating water for the conversion of HCO_3^- to formic acid with Fe is postulated in Figure 5.15. Initially, Fe is oxidized into Fe_3O_4 in water with the help of HCO_3^- to produce a large amount of hydrogen, which simultaneously reduce in situ Fe_3O_4 to lead to the more active sites on the surface of Fe_3O_4 (Fe_3O_{4-x}). Subsequently, the formed hydrogen and HCO_3^- are activated by adsorbing on the Fe_3O_{4-x} surface, and then the activated H on the Fe_3O_{4-x} surface attacks C=O, followed by the hydroxyl group of HCO_3^- leaving. Finally, formate is formed along with H_2O. Although detailed studies are needed, the proposed autocatalytic role of Fe_3O_{4-x} formed under hydrothermal reducing conditions in this study may provide some useful experimental data for understanding abiotic organic synthesis at deep-sea hydrothermal vents.

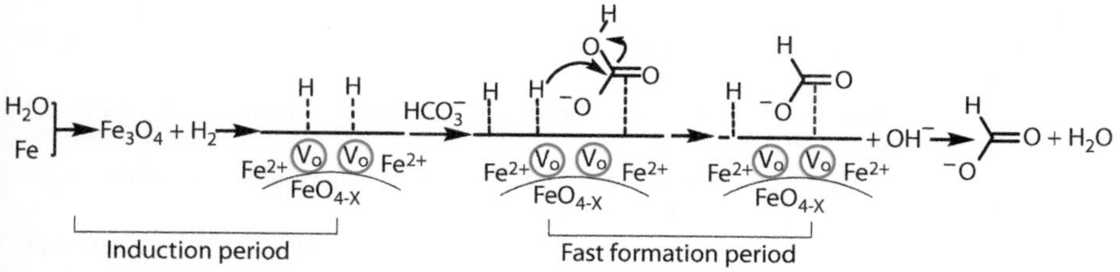

FIGURE 5.15 Proposed mechanism of reduction of HCO_3^- into formate with Fe.

5.4 CONCLUSIONS

In summary, highly efficient water splitting for CO_2 reduction into formic acid with a general Fe powder in hot water without adding any other catalyst was achieved, which obtained an excellent formic acid yield of 92%. In this process, we found that HCO_3^- plays a key role in enhancing Fe oxidation in water to produce hydrogen, and the mechanism of autocatalytic reduction CO_2 involves the formation of Fe_3O_{4-x} by in situ reducing Fe_3O_4 with the formed hydrogen. The exact mechanism of this water splitting for CO_2 reduction is under further investigation. This work not only demonstrates a promising method to highly efficiently reduce atmospheric CO_2 linked to global climate change into organic chemicals but also provides some useful experimental data to further understand the abiotic organic synthesis at deep-sea hydrothermal vents.

REFERENCES

Anderson, J., Kuhn, M. and Diebold, U. 1996. Epitaxially grown Fe_3O_4 thin films: An XPS study. *Surf. Sci. Spectra* 4 (3):266–272.

Boddien, A., Loges, B., Gärtner, F. et al. 2010. Iron-catalyzed hydrogen production from formic acid. *J. Am. Chem. Soc.* 132 (26):8924–8934.

Chueh, W. C. and Haile, S. M. 2009. Ceria as a thermochemical reaction medium for selectively generating syngas or methane from H_2O and CO_2. *ChemSusChem*. 2 (8):735–739.

Franck, E. 1983. Thermophysical properties of supercritical fluids with special consideration of aqueous systems. *Fluid Phase Equilibr.* 10 (2):211–222.

Galvez, M., Loutzenhiser, P., Hischier, I. and Steinfeld, A. 2008. CO_2 splitting via two-step solar thermochemical cycles with Zn/ZnO and FeO/Fe_3O_4 redox reactions: Thermodynamic analysis. *Energy Fuels* 22 (5):3544–3550.

Hawn, D. D. and DeKoven B. M. 1987. Deconvolution as a correction for photoelectron inelastic energy losses in the core level XPS spectra of iron oxides. *Surf. Interface Anal.* 10 (2–3):63–74.

Horita, J. and Berndt, M. E. 1999. Abiogenic methane formation and isotopic fractionation under hydrothermal conditions. *Science* 285 (5430):1055–1057.

Hu, Y. H. 2012. A highly efficient photocatalyst—Hydrogenated black TiO_2 for the photocatalytic splitting of water. *Angew. Chem. Int. Ed.* 51 (50):12410–12412.

Huber, G. W., Shabaker, J. and Dumesic, J. 2003. Raney Ni-Sn catalyst for H production from biomass-derived hydrocarbons. *Science* 300 (5628):2075–2077.

Jin, F., Gao, Y., Jin, Y. J. et al. 2011. High-yield reduction of carbon dioxide into formic acid by zero-valent metal/metal oxide redox cycles. *Energy Environ. Sci.* 4 (3):881–884.

Jin, F., Kishita, A., Moriya, T. and Enomoto, H. 2001. Kinetics of oxidation of food wastes with H_2O_2 in supercritical water. *J. Supercrit. Fluids* 19 (3):251–262.

Jin, F., Zhou, Z., Moriya, T. et al. 2005. Controlling hydrothermal reaction pathways to improve acetic acid production from carbohydrate biomass. *Environ. Sci. Technol.* 39 (6):1893–1902.

Lane, N. and Martin, W. F. 2012. The origin of membrane bioenergetics. *Cell* 151 (7):1406–1416.

Martin, W., Baross, J., Kelley, D. and Russell, M. J. 2008. Hydrothermal vents and the origin of life. *Nat. Rev. Microbiol.* 6 (11):805–814.

Maurice, V., Yang, W. and Marcus, P. 1996. XPS and STM study of passive films formed on Fe–22Cr (110) single-crystal surfaces. *J. Electrochem. Soc.* 143 (4):1182–1200.

McCammon, C. 2005. The paradox of mantle redox. *Science.* 308 (5723):807–808.

McCollom, T. M. and Seewald, J. S. 2001. A reassessment of the potential for reduction of dissolved CO_2 to hydrocarbons during serpentinization of olivine. *Geochim. et Cosmochim. Acta* 65 (21):3769–3778.

Moret, S., Dyson, P. J. and Laurenczy, G. 2014. Direct synthesis of formic acid from carbon dioxide by hydrogenation in acidic media. *Nat. Commun.* 5:1–7.

Muhler, M., Schlögl, R. and Ertl, G. 1992. The nature of the iron oxide-based catalyst for dehydrogenation of ethylbenzene to styrene 2. Surface chemistry of the active phase. *J. Catal.* 138 (2):413–444.

Oblonsky, L. and Devine, T. 1995. A surface enhanced Raman spectroscopic study of the passive films formed in borate buffer on iron, nickel, chromium and stainless steel. *Corros. Sci.* 37 (1):17–41.

Ogundele, G. and White, W. 1986. Some observations on corrosion of carbon steel in aqueous environments containing carbon dioxide. *Corrosion* 42 (2):71–78.

Shebanova, O. N. and Lazor, P. 2003. Raman study of magnetite (Fe_3O_4): Laser-induced thermal effects and oxidation. *J. Raman Spectrosc.* 34 (11):845–852.

Uhm, S., Chung, S. T. and Lee, J. 2008. Characterization of direct formic acid fuel cells by impedance studies: In comparison of direct methanol fuel cells. *J. Power Sources* 178 (1):34–43.

Webb, T., Kruger, J. and MacKenzie, R. 1970. Differential thermal analysis. *Fundamental Aspects* 1:327.

Wu, B., Gao, Y., Jin, F. et al. 2009. Catalytic conversion of $NaHCO_3$ into formic acid in mild hydrothermal conditions for CO_2 utilization. *Catal. Today* 148 (3–4):405–410.

Yadav, R. K., Baeg, J. O., Oh, G. H. et al. 2012. A photocatalyst–enzyme coupled artificial photosynthesis system for solar energy in production of formic acid from CO_2. *J. Am. Chem. Soc.* 134 (28):11455–11461.

Yamashita, T. and Hayes, P. 2008. Analysis of XPS spectra of Fe^{2+} and Fe^{3+} ions in oxide materials. *Appl. Surf. Sci.* 254 (8):2441–2449.

Yang. Z. Z., He, N. L., Zhao, Y. N., Li, B. and Yu, B. 2011. CO_2 capture and activation by superbase/polyethylene glycol and its subsequent conversion. *Energy Environ. Sci.* 4 (10):3971–3975.

Zhao, Y. N., Yang. Z. Z., Luo, S. H. and He, N. L. 2013. Design of task-specific ionic liquids for catalytic conversion of CO_2 with aziridines under mild conditions. *Catal. Today* 200:2–8.

6 Hydrothermal Reduction of CO_2 to Low-Carbon Compounds

Ge Tian, Chao He, Ziwei Liu, and Shouhua Feng

CONTENTS

6.1 Hydrothermal Reactions from Carbon Dioxide to Phenol 79
6.2 Hydrothermal Reactions from Carbon Dioxide to Simple Carboxylic Acids ... 81
6.3 Hydrothermal Reactions from Carbon Dioxide to Methane 85
References 88

The reduction of CO_2 under hydrothermal conditions has attracted more attention in recent years [1]. On the one hand, it is a feasible program to solve the "Greenhouse Effect" problem; on the other hand, it has plausible implications for the abiotic synthesis of complex organic molecules in prebiotic chemistry because the hydrothermal system is considered to be one of the most suitable environments for the origin of life [2,3].

Furthermore, iron, which accounts for 5.0 wt% of the Earth's crust, is abundant and widespread. It is widely used as a catalyst in the reduction of CO_2 and could play an important role in prebiotic synthesis.

6.1 HYDROTHERMAL REACTIONS FROM CARBON DIOXIDE TO PHENOL [4]

In previous studies, hydrothermal conversion of carbonates into phenol was reported [5]. Because the research about the chemical utilization of carbon dioxide is significant for environmental purposes, the current study aims to confirm that carbon dioxide gas follows the mechanism of carbonate conversion and determine if carbon dioxide gas reacts with water under hydrothermal conditions. The result shows that carbon dioxide gas can be converted into phenol by iron powder as ecofriendly catalysts and under mild hydrothermal conditions.

The study is based on well-established hydrothermal methods [6]. Purified carbon dioxide gas, water, and metal catalyst powders were put in a reactor. Gas chromatography–mass spectroscopy (GC-MS) results indicated that no organic compound could be detected in the purified carbon dioxide gas starting material (99.999%). The metal powders, water, and reactor system used in the experiment were purified according to the treatment procedures reported earlier and free of organic contamination [7]. In a run, 560 mg (10 mm) of purified iron powder was mixed with water in a glass-liner Endeavor Catalyst Screening System (ECSS) with a filling capacity of 50% in a total volume of 20 mL. Carbon dioxide gas was injected into the ECSS, increasing the inside pressure to 1.4 MPa. The system was heated at 200°C for 120 h. The final pressure and pH of the system after cooling to room temperature were 1.0 MPa and 5.5, respectively. Product analysis was carried out with GC-MS. A mass spectrum of the solution after reaction showed the formation of phenol, consistent with its standard mass spectrum, according to

$$CO_2 + H_2O \rightarrow C_6H_5OH. \tag{6.1}$$

We tested the change of the iron powder after the reaction. After reaction, 1 wt% Fe^{2+} was proven to be present in the solution by inductively coupled plasma analysis. Thus, a possible oxidization mechanism of iron under the hydrothermal conditions is proposed:

$$Fe + H_2O \rightarrow Fe^{2+} + H_2 \tag{6.2}$$

Apart from iron powder, other metal powders and zeolites were used under the same reaction conditions. These materials, including Al, Zn, Co, and Ni powders, as well as Fe_3O_4, Fe^{2+}, Fe^{3+}, Co^{2+}, Ni^{2+}-modified ZSM-5, MCM-41, MCM-48, H-beta, and Na-Y zeolites, did not form significant phenol formation; only trace amounts of phenol were obtained using Co and Ni powders as catalyst. The possible role of oxygen in the reaction was studied and both the products and the yield remained unchanged with changing oxygen concentration.

Kinetic studies were carried out in the system. Figure 6.1 shows a kinetic curve for the reaction. The yield was obtained by comparing a calibration curve of peak area with the peak area of the product. In the first 10 h of the reaction, the yield of phenol

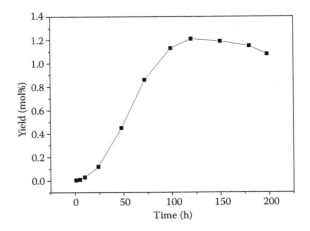

FIGURE 6.1 Kinetic curve for the hydrothermal production of phenol.

increased slowly. With the increasing of reaction time from 10 to 120 h, the phenol yield increased rapidly. This is typical of a self-catalytic reaction, with a fast formation reaction after a short induction period. In the short induction period, some compounds that can further catalyze the reaction (such as formaldehyde, formic acid, and others) are formed. At this stage, several organic molecules could be identified by GC-MS, including the main product phenol and trace amounts of formic acid and formaldehyde. After 120 h, the phenol yield reached a maximum value of 1.21 mol% according to carbon dioxide and then remained nearly constant. The final product was phenol.

In the past decade, organic syntheses starting from carbon dioxide have also been widely discussed [8,9]. Using Fe_3O_4 as catalyst, the production of phenol from carbon dioxide reduced in supercritical state was reported [10]. In comparison, the hydrothermal approach is highly selective under lower temperature and pressure. As an insight into the origin of life, it would be important to prove that carbon dioxide can react with water under hydrothermal conditions. In this process, dissolved carbon dioxide chemisorbs onto the surface of iron atoms, and the reduction of carbon dioxide is closely related to the oxidation of iron atoms [5]. Therefore, we think that the process includes a catalysis reaction and a redox reaction of carbon dioxide. The reactions take place much more efficiently with the use of iron powder than with other powders, under the same reaction conditions. This is consistent with the relatively large redox potential of iron in group VIIIB: Fe, 0.44 V; Co, 0.277 V; and Ni, 0.25 V. Because iron is the most abundant metallic, redox-active element on Earth, the iron powder catalyst is ecofriendly, widespread, and easily prepared.

Based on the observation of the final product phenol and intermediate formaldehyde and formic acid in the hydrothermal reactions, we conclude that the reaction mechanism from carbon dioxide to phenol does follow the mechanism proposed previously (Figure 6.2) [5]. At the first stage of the reaction, the iron metal reacts with water to form H_2 and Fe^{2+}. Meanwhile, the dissolved CO_2 molecules are adsorbed onto the surface of the iron powder. Through attack of H_2, CO_2 is reduced to produce formaldehyde and formic acid. The complicated processes involve two simple types of reactions: oxidative coupling and rearrangement reactions. The two reactions proceed alternately and the final product phenol was produced. In the reaction, as a source of hydrogen, water is necessary, whereas the ratio of carbon dioxide to water has little effect on the reduction of carbon dioxide. The conversion of carbon dioxide into phenol becomes an alternative to the existing electrochemical and biotic techniques to reuse the greenhouse gas carbon dioxide. The decreasing carbon dioxide pressure and accumulation of the product phenol limit the further conversion of carbon dioxide into phenol. Therefore, we carried out a constant-pressure experiment by continuously supplying enough carbon dioxide, and the results showed an increasing tendency for the phenol yield. By gradually collecting phenol, we believe that this simple process can be repeated to reach an industrial level.

6.2 HYDROTHERMAL REACTIONS FROM CARBON DIOXIDE TO SIMPLE CARBOXYLIC ACIDS [11]

The reactions of CO_2 with water in the presence of iron nanoparticles were carried out using the ECSS (Argonaut Technologies, Inc.). The result of GC-MS showed that

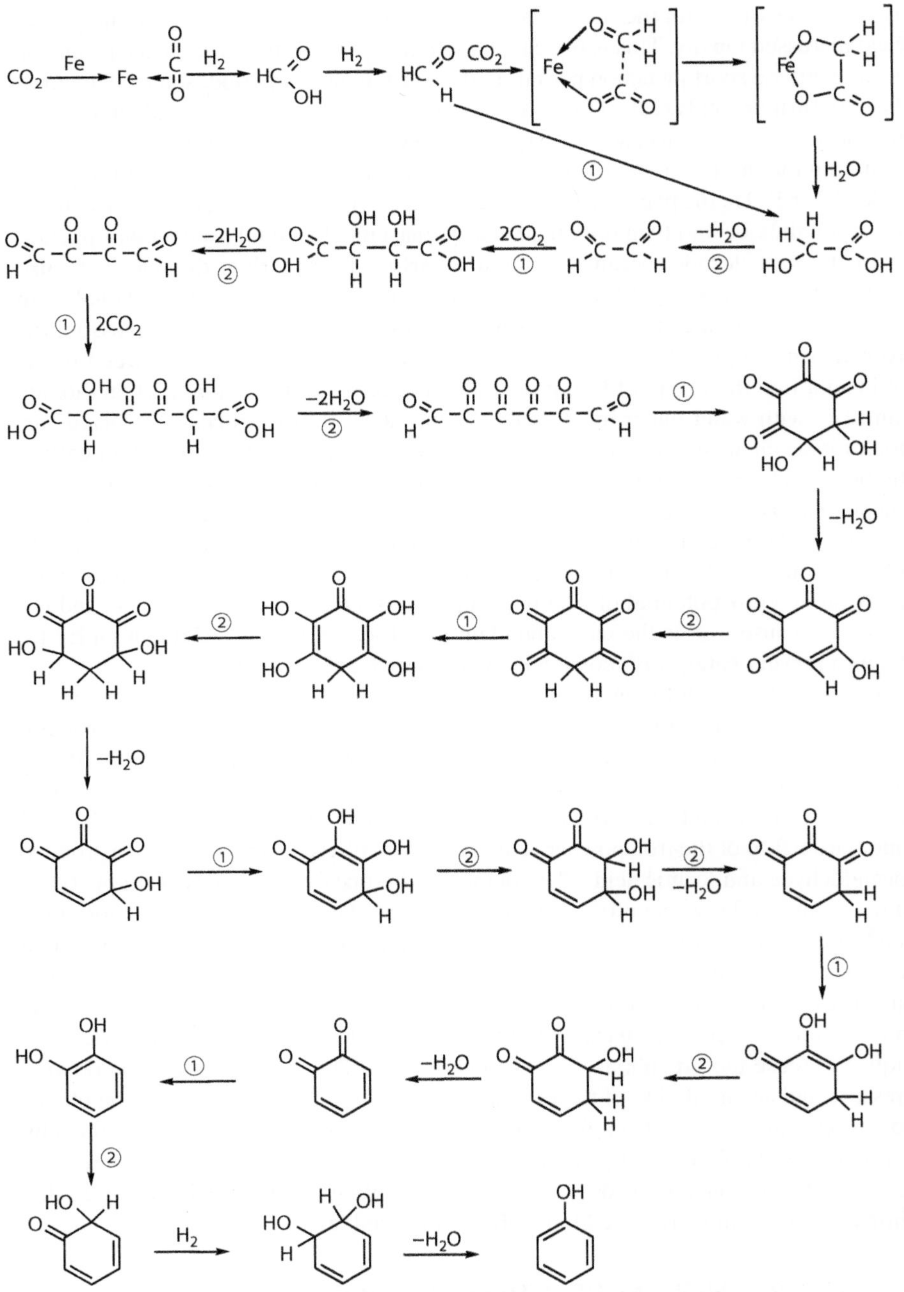

FIGURE 6.2 Formation mechanisms of phenol under hydrothermal conditions. (① refers to the oxidative coupling reaction; ② refers to the rearrangement reaction.)

Hydrothermal Reduction of CO_2 to Low-Carbon Compounds

CO_2 (99.999%) and ultrapure water were no organic contamination. The catalyst, iron nanoparticles, was produced using reducing ferrous ion. The reactor was sealed and pressurized to 0.14–1.4 MPa by CO_2 gas after putting the catalyst and ultrapure water into it. The reactor was subjected to a hydrothermal treatment at 80–200°C for 5–200 h. After the reaction, the pH value of the mixture was 5.5. The product was identified by GC-MS.

After the reaction, formic acid and acetic acid were produced when the mass spectra of the products were compared to that of standard substance. As shown in Figure 6.3a and b, the yields of formic acid and acetic acid increase with the increase of reaction time, reaching the maximum value after 72 h, and then remain nearly constant. The yields of formic acid and acetic acid are 8.5 and 3.5 mmol·L^{-1}, respectively.

Higher CO_2 pressure can increase the yield of formic acid because a more active center is formed at the surface of the iron nanoparticles with more CO_2 adsorbed. The volume of water has little effect on the conversion of CO_2.

The effect of temperature was also taken into account. As illustrated in Figure 6.3c, the yield of formic acid increases markedly with the increase of temperature, while the yield of acetic acid increases less, and the total conversion of CO_2 increases as the temperature increases. We think that more and faster H_2 is produced at higher temperature, which provides a stronger reducing environment and promotes the conversion of CO_2.

On the basis of the above result, we proposed a possible reaction mechanism of formic acid and acetic acid formation. Figure 6.4 shows the main process of the reaction. At the first stage of the reactions, the iron nanoparticles reacted with water to produce H_2. Meanwhile, the dissolved CO_2 molecule in water was adsorbed at the surface of the iron nanoparticles. After the attack of H_2, CO_2 was reduced to intermediate A, which was hydrolyzed to formic acid. Intermediate A continued to react with H_2 to generate intermediate B. At last, intermediate A and intermediate B reacted to form intermediate C, which was hydrolyzed to form acetic acid.

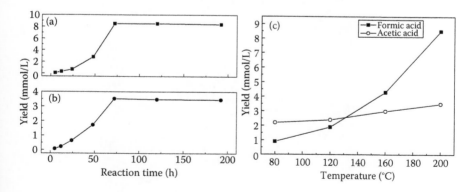

FIGURE 6.3 Kinetic curve of the hydrothermal production of formic acid (a) and acetic acid (b). Water: 6 ml, CO_2: 1.4 MPa, Fe: 5 mmol, temperature: 200°C. Yields of formic acid and acetic acid at different temperature (c). Water: 6 ml, CO_2: 1.4 MPa, Fe: 5 mmol, reaction time: 72 h.

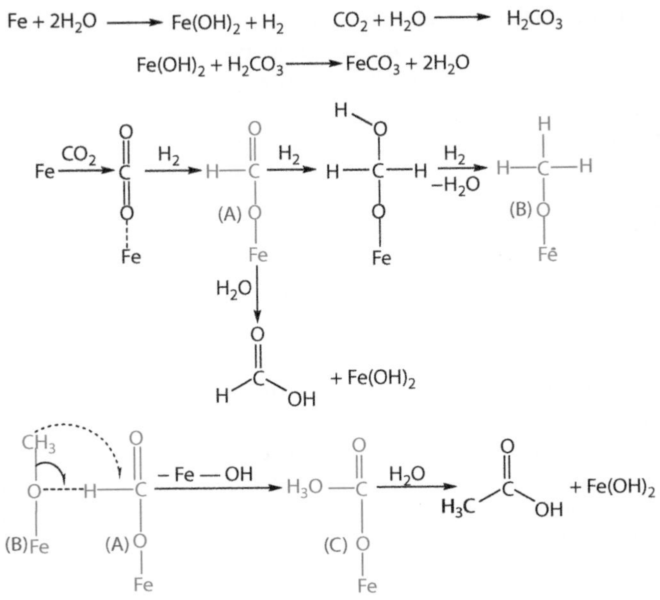
Fe + 2H2O → Fe(OH)2 + H2
CO2 + H2O → H2CO3
Fe(OH)2 + H2CO3 → FeCO3 + 2H2O
CO2
H2
H2
H2
−H2O
(A)
(B)
H2O
+ Fe(OH)2
− Fe — OH
H2O
(C)
+ Fe(OH)2

FIGURE 6.4 Proposed mechanism for the formation of formic acid and acetic acid in the presence of nano-Fe under hydrothermal conditions.

In the reaction system, the iron nanoparticles act as a reducing agent to catalyze the reduction of CO_2. As shown in Figure 6.5, the diameter of the particles was about 200 nm. The iron nanoparticles reacted rapidly with water to form hydrogen, whose formation was confirmed by igniting. After the reaction, the unreacted iron was separated using a magnet, and the residual solid was collected to analyze by powder

General
SEI 5.0kV ×30,000 100nm WD 5.7 mm

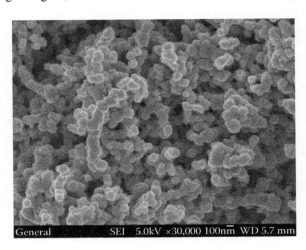

FIGURE 6.5 The low-magnification scanning electron microscopy images of the iron nanoparticles.

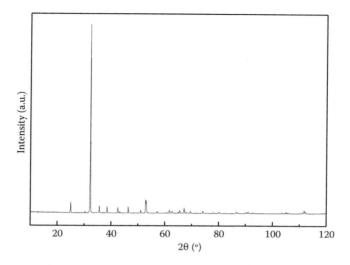

FIGURE 6.6 XRD pattern of the residual solid after removing unreacted iron using magnet.

x-ray diffraction (XRD). The result showed that the solid was ferrous carbonate (Figure 6.6), concluding that a great part of the iron nanoparticles was converted to ferrous carbonate after the reaction. The production of H_2 and ferrous carbonate also confirmed the suggested mechanism.

The products of the reaction are formic acid and acetic acid instead of phenol obtained in our previous work. It was deduced that the amount of rapid formation of H_2 plays an important role in the selectivity of the product. At the beginning of the reaction, more H_2 was generated from the reaction of the iron nanoparticles with water than from the reaction of iron powder with water, which provided a stronger reducing environment. Each adsorbed carbon dioxide molecule formed an active reaction center, which was reduced by hydrogen to form formic acid and acetic acid under the environment.

6.3 HYDROTHERMAL REACTIONS FROM CARBON DIOXIDE TO METHANE [12]

A feasible method is developed in which carbon dioxide is directly reduced to some organic compounds such as phenol, formaldehyde, and carboxylic acids in hydrothermal systems. It is noticed that all of the materials possess a small size effect, even if the change is in a small range [13–16]. Different products are reported with the size of the catalyst ranging from about 75 μm to about 200 nm. We present here a further study on the hydrothermal reduction of carbon dioxide in the presence of much smaller-sized iron nanoparticles. The reaction obtains the single product methane, rather than as a part of products in previously reported systems [17–19], and uses carbon dioxide as a raw material rather than additional H_2 as in the case for the Al_2O_3-based catalysts [20,21].

The study on the reactions of carbon dioxide with water in the presence of iron nanoparticles was also based on well-established hydrothermal methods. The reactions started with purifying carbon dioxide gas, water, and metal catalyst powder. CO_2 (99.999%) and ultrapure water were free of any organic contamination checking by GC-MS. Iron nanoparticles with an average diameter of 100 nm were prepared by reducing ferrous ion. The reaction of CO_2 with water using iron powder as catalyst was carried out in autoclaves. Iron nanoparticles (5 mmol) and ultrapure water (5 mL) were put into the reactor and sealed, and then the reactor was pressurized by CO_2 up to 0.14–1.4 MPa. All substrates underwent mild hydrothermal treatments at 80–200°C for 5–200 h. After the reactions, the system was cooled down to ambient temperature and gaseous products were collected. The final pressure and pH value were ca. 0.1–1.0 MPa and 5.5, respectively. Gaseous- and liquid-phase products were analyzed by GC-MS.

In gaseous phase, CH_4 was formed, while the detection of the organic in liquid was below the limitation. The yield of CH_4 increased with an increase in reaction time from 5 to 72 h, as shown in Figure 6.7a. After 72 h, the yield reached a maximum value at 1.96 mol% and then remained nearly constant. The effect of temperature was also taken into account. The higher the reaction temperature, the more CO_2 was converted to CH_4, as shown in Figure 6.7b. Obviously, the higher activity CO_2 was easily reduced at higher temperatures. The higher pressure corresponded to more adsorbed CO_2 on the surface of the iron nanoparticles, and the higher pressure was also beneficial for the reduction of CO_2.

We have also given the possible reaction mechanism for the formation of methane. Similar to previously suggested formation mechanisms for phenol and carboxylic acids, the iron nanoparticles acted as both reducing and catalytic agents. Part of the iron nanoparticles reacted with water to generate H_2, and simultaneously, part of the iron nanoparticles absorbed and reduced the dissolved CO_2 molecules by the attack of H_2 on and at the surface of iron nanoparticles. Methane was then formed. In order to study the catalytic effect of the iron nanoparticles, we conducted comparison experiments; for example, H_2 gas as a feedstock was introduced into the

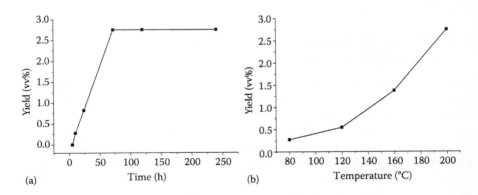

FIGURE 6.7 (a) Kinetic curve of the hydrothermal production of CH_4. (b) Yields of CH_4 at different temperatures.

CO_2–H_2O system without iron nanoparticles. According to the reaction mechanism mentioned above, methane was not detected in the gaseous phase or the liquid phase. The formation reaction for methane is as indicated in Figure 6.8.

The average diameter of the residual solids is the same before and after the reactions (Figure 6.9a and b), but they are a mixture of iron and ferrous carbonate. The iron powder was separated by magnet, and the residual solids were identified as ferrous carbonate by powder XRD. It was apparent that water and iron powder reacted with CO_2 to form ferrous carbonate under hydrothermal conditions. To prove the catalysis effect of the iron nanoparticles, we carried out two experiments where one was without any catalysts in the system and the other was in the presence of ferrous carbonate acting as catalyst under the same conditions. The results showed that there

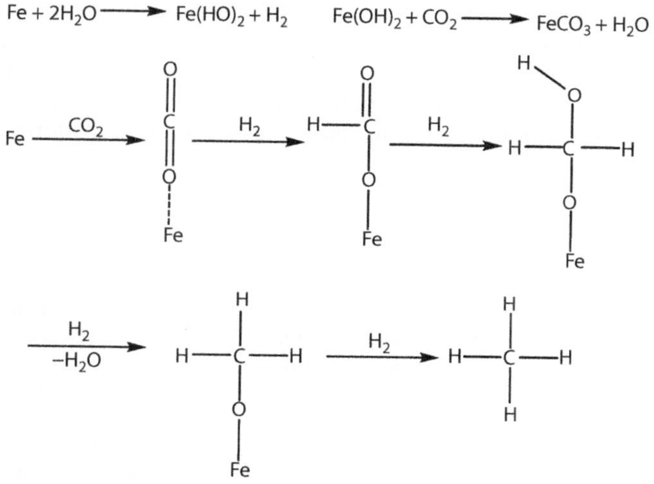

FIGURE 6.8 Proposed mechanism of methane formation.

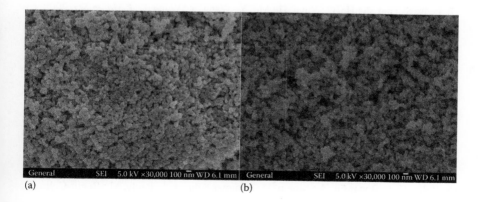

(a)   (b)

FIGURE 6.9 Low-magnification scanning electron microscopy images of a 100-nm iron nanoparticle (a) before and (b) after hydrothermal reactions.

were no organic compounds in the gaseous phase or in the liquid phase in the above two experiments.

In summary, we have conducted mild hydrothermal reactions from CO_2 to phenol, formic acid and acetic acid, and CH_4 in the presence of iron powder about 75 μm, 200 nm, and 100 nm in diameter, respectively, and have characterized the reaction process. The ready CO_2 conversion process follows the main principles of green chemistry (e.g., one-step process under mild conditions, ecofriendly catalyst, no toxicity to human health and environment, and no organic solvents). The work will help in the utilization of CO_2 in the future.

Compared to other reported methods for the reduction of CO_2, the characteristics of this reaction system are as follows: first, the catalyst, Fe powder, can be prepared readily, with iron being a widespread and abundant element forming the Earth's crust, as well as much of our Earth's outer and inner core. Second, the reaction is carried out under lower pressure and temperature, which is readily accessible. Third, additional H_2 as a reducing agent does not need to be added to the reaction system because H_2 is autogenously generated in the reaction. Fourth, water as the source of hydrogen is necessary in this reaction, and it does not impede the formation of the organic matter.

REFERENCES

1. (a) Horita, J.; Berndt, M. E. *Science* 1999, 285, 1055. (b) Takahashi, H.; Kori, T.; Onoki, T.; Tohji, K.; Yamasaki, N. *J. Mater. Sci.* 2008, 43, 2487. (c) Takahashi, H.; Liu, L. H.; Yashiro, Y.; Ioku, K.; Bignall, G.; Yamasaki, N. *J. Mater. Sci.* 2006, 41, 1585.
2. Corliss, J. B.; Dymond, J.; Gordon, L. I.; Edmond, J. M.; von Herzen, R. P.; Ballard, R. D.; Green, K.; Williams, D.; Bainbridge, A.; Crane, K.; van Andel, T. H. *Science* 1979, 203, 1073.
3. Martin, W.; Baross, J.; Kelley, D.; Russell, M. J. *Nat. Rev.* 2008, 6, 805.
4. Tian, G.; He, C.; Chen, Y.; Yuan, H.; Liu, Z.; Shi, Z.; Feng, S. *ChemSusChem* 2010, 3, 323.
5. Tian, G.; Yuan, H.; Mu, Y.; He, C.; Feng, S. *Org. Lett.* 2007, 9, 2019.
6. Feng, S.; Xu, R. *Acc. Chem. Res.* 2001, 34, 239.
7. Feng, S; Tian, G.; He, C.; Yuan, H.; Mu, Y.; Wang, Y.; Wang, L. *J. Mater. Sci.* 2008, 43, 2418.
8. Matsuo, T.; Kawaguchi, H. *J. Am. Chem. Soc.* 2006, 128, 12362.
9. Chen, Q.; Bahnemann, D. *J. Am. Chem. Soc.* 2000, 122, 970.
10. Chen, Q.; Qian, Y. *Chem. Commun.* 2001, 1402.
11. He, C.; Tian, G.; Liu, Z.; Feng, S. *Org. Lett.* 2010, 12, 649.
12. Liu, Z.; Tian, G.; Zhu, S.; He, C.; Yue, H.; Feng, S. *ACS Sustainable Chem. Eng.* 2013, 1, 313.
13. Martinez-Perez, M. J.; de, M. R.; Carbonera, C.; Martinez-Julvez, M.; Lostao, A.; Piquer, C.; Gomez-Moreno, C.; Bartolome, J.; Luis, F. *Nanotechnology* 2010, 21 (46), 465707/1.
14. Guisbiers, G. *Nanoscale Res. Lett.* 2010, 5 (7), 1132.
15. Lian, J.; Liang, Y.; Kwong, F.-l.; Ding, Z.; Ng, D. H. L. *Mater. Lett.* 2012, 66 (1), 318.
16. Belle, C. J.; Bonamin, A.; Simon, U.; Santoyo-Salazar, J.; Pauly, M.; Begin-Colin, S.; Pourroy, G. *Sens. Actuators, B* 2011, 160 (1), 942.
17. Pan, J.; Wu, X.; Wang, L.; Liu, G.; Lu, G. Q.; Cheng, H.-M., *Chem. Commun.* 2011, 47 (29), 8361.

18. Xi, G.; Ouyang, S.; Ye, J. *Chem. Eur. J.* 2011, 17 (33), 9057.
19. Takahashi, H.; Kori, T.; Onoki, T.; Tohji, K.; Yamasaki, N. *J. Mater. Sci.* 2008, 43 (7), 2487.
20. Jacquemin, M.; Beuls, A.; Ruiz, P. *Catal. Today* 2010, 157 (2010), 462.
21. Gnanamani, M. K.; Shafer, W. D.; Sparks, D. E.; Davis, B. H. *Catal. Commun.* 2011, 12 (11), 936.

7 Hydrothermal CO$_2$ Reduction with Zinc to Produce Formic Acid

Yang Yang, Guodong Yao, Binbin Jin, Runtian He, Fangming Jin, and Heng Zhong

CONTENTS

7.1 Introduction 91
7.2 Experimental Section 93
    7.2.1 Materials 93
    7.2.2 Experimental Procedure 93
    7.2.3 Product Analysis 93
7.3 Formate Production with NaHCO$_3$ and Zn under Hydrothermal Conditions ... 94
    7.3.1 The Reduction of NaHCO$_3$ with Zn under Hydrothermal Conditions ... 94
    7.3.2 The Reaction Characteristics and the Optimization of the Reaction Parameters 94
7.4 Water Splitting for Hydrogen Production with Zn 96
7.5 Formic Acid Production with Gaseous CO$_2$ 97
7.6 Proposed Mechanism for Formate Production 99
7.7 Quantum Chemical Calculations of the Reaction Mechanism 101
7.8 Zn–ZnO Cycle and Assessment of the Energy Conversion Efficiency 105
7.9 Conclusions 106
Acknowledgments 106
References 106

7.1 INTRODUCTION

Artificial photosynthesis, which converts solar energy into chemical energy, has attracted research attention extensively, as it solves the problem of energy crisis thoroughly. Despite the great effort paid in this area,[1–4] high-efficiency conversion with direct use of solar energy remains challenging. In contrast to the direct utilization of solar energy, it has been reported that a high efficiency of artificial photosynthesis can be realized through integrated technology, which uses metals for water splitting and solar energy for metal oxide reduction, for example, the solar-driven two-step water-splitting thermochemical cycle based on metal/metal oxide redox reactions for hydrogen production,[5–10] and the dissociation of CO$_2$ and H$_2$O into chemical fuel by using certain two-step solar-driven redox reactions.[11,12] However, the product of CO$_2$

was limited to CO, and the produced hydrogen cannot be used for CO_2 conversion directly.

On the other hand, hydrothermal reactions have shown great potential in splitting water to produce hydrogen; for example, geological studies showed that Fe-bearing minerals in mantle could produce hydrogen under hydrothermal conditions.[13] Further, hydrothermal reactions have played an important role in the formation of fossil fuels, like the abiotic conversion of dissolved CO_2 into hydrocarbons in the Earth's crust,[13–15] in which CO_2 was reduced by the minerals generated by hydrogen and mantle minerals acted as the catalysts as well. The abiotic synthesis of organics suggests that highly efficient dissociation of H_2O and subsequent reduction of CO_2 into organics could be achieved with metals under hydrothermal conditions.

Although metals are oxidized for water splitting and CO_2 reduction under hydrothermal conditions, metal oxide can be reduced by concentrated solar energy.[5–10] Thus, an integrated technology of interest for high-efficiency artificial photosynthesis could be developed by coupling the geochemical reactions involved in the dissociation of H_2O and the reduction of CO_2 in the presence of metals with the solar-driven thermochemical reduction of metal oxides into metals, as described in Figure 7.1.

With this concept, the high-efficiency reduction of CO_2 by metals is crucial because the reduction of metal oxides with solar energy has been well studied and is nearing practical application.[5–10] Among the candidate metals that have potential for reducing water into hydrogen in the hydrothermal reduction of CO_2, Zn has a strong thermodynamic driving force for oxide formation, and ZnO is often used as a catalyst in the catalytic reduction of CO_2 with hydrogen; thus, the ZnO formed in situ by Zn oxidation in water might have an autocatalytic role and lead to a simple and highly efficient reduction of CO_2 without the need for complex material microstructure design. Therefore, Zn is selected as the reducing agent for the autocatalytic hydrothermal CO_2 reduction.

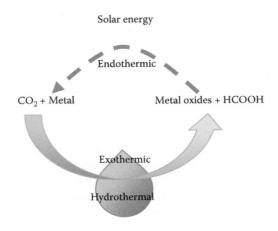

FIGURE 7.1 Proposed integrated technology for formic acid production.

Hydrothermal CO_2 Reduction with Zinc to Produce Formic Acid

In this chapter, we summarize the recent advance in highly efficient H_2O dissociation with Zn for CO_2 reduction into formic acid. The overall reaction can be expressed in Equation 7.1, in which the Gibbs free energy ($\Delta G°$) and the heat of reaction ($\Delta H°$) were calculated using available thermodynamic data:

$$Zn + CO_2 + H_2O = HCOOH + ZnO$$
$$\Delta G°(298\ K) = -23.01\ kJ/mol\ \Delta H°(298\ K) = -114.75\ kJ/mol \quad (7.1)$$

Both ΔG and ΔH values are negative, which means that this reaction does not require extra energy input and can be potentially self-supported. As Zn can be readily reduced to its zero-valent state by using solar energy,[8,16] the proposed strategy can be regarded as a method for converting solar energy into fuel. The content mainly includes (1) formic acid syntheses from CO_2 with water-derived hydrogen, (2) hydrogen production from water with Zn, (3) reaction mechanism study and density function theory calculation for formic acid synthesis from CO_2 and water with Zn, and (4) solar to chemical energy conversion efficiency estimation.

7.2 EXPERIMENTAL SECTION

7.2.1 Materials

Zn powder (325 mesh, ≥98% from Alfa Aesa) was used in this study. $NaHCO_3$ (AR, ≥98% from Sinopharm Chemical Reagent Co., Ltd) was used as a CO_2 resource to simplify handling. Gaseous CO_2 and H_2 (>99.995%) were purchased from Shanghai Poly-Gas Technology Co., Ltd. In this study, deionized water was used in all experiments.

7.2.2 Experimental Procedure

Stainless steel 316 tubing [9.525 mm (3/8 in) o.d., 1 mm wall thickness and 120 mm length] reactors, providing the inner volume of 5.7 mL, is used in the experiments. A brief experiment procedure is given below. The desired amount of $NaHCO_3$ (CO_2 source), reductant (Zn powder), and deionized water was loaded in the batch reactor. After loading, the reactor was immersed in a salt bath at a preset temperature. During the reaction, the reactor was shaken and kept horizontally to enhance the mixture and heat transfer. After the preset reaction time, the reactor was removed from the salt bath to quench in a cold-water bath. After the reactions, the liquid, gaseous, and solid samples were collected for analysis. The reaction time was defined as the duration of time that the reactor was kept in the salt bath.

7.2.3 Product Analysis

Liquid samples were filtered (0.22 μm filter film) and then analyzed by high-performance liquid chromatography (HPLC, Aglient 1260), total organic carbon (TOC, Shimadzu TOC 5000A), and gas chromatography/mass spectroscopy (GC/MS, Agilent 7890).

The solid samples were washed with deionized water three times to remove impurities and with ethanol three times to make the solid sample dry quickly. The samples were then dried in an isothermal oven at 40°C for 3–5 h and characterized using x-ray diffraction (XRD, Shimadzu XRD-6100), scanning electron microscopy (SEM, FEI Quanta 200), and Fourier transform infrared (FT-IR) spectroscopy.

7.3 FORMATE PRODUCTION WITH $NaHCO_3$ AND Zn UNDER HYDROTHERMAL CONDITIONS

7.3.1 The Reduction of $NaHCO_3$ with Zn under Hydrothermal Conditions

The investigation started with the reaction of $NaHCO_3$ and Zn without the addition of any catalyst. $NaHCO_3$ was used as the source of CO_2 because CO_2 not only can be readily absorbed in NaOH solutions to form $NaHCO_3$ but also can promote the hydrogen production from water as discussed later. The reduction of $NaHCO_3$ and Zn under hydrothermal conditions was performed over a wide range of reaction conditions. All the reaction products in the liquid samples were pure, and only formate was observed by HPLC, GC, and IC analyses. Gas samples analyzed by GC/TCD showed that hydrogen, a small amount of CO_2, and a trace amount of CH_4 were produced when reaction times were longer than 60 min. Organic carbon in the liquid samples was determined by TOC, and the amount of carbon in formate was comparable to the total carbon in the samples, indicating that the selectivity for the production of formate was approximately 100%. These results indicate that formate can be easily and selectively produced by $NaHCO_3$ reacting with Zn under hydrothermal conditions.

XRD analysis showed that the oxidation of Zn was rapid; as shown in Figure 7.2a and b, almost all the Zn was oxidized to ZnO after only 10 min. The morphology of Zn powder and the solid products after the reaction was characterized by SEM (Figure 7.2c and d). The reagent Zn powder before reactions was an aggregation of many small particles and had an irregular shape. Zn powder was used for the reaction without any pretreatment. However, the solid products after reactions were small flower-like tetrapod whiskers with a size of less than 10 μm, which is the typical crystalline form of ZnO, further indicating that Zn powder was converted to ZnO after the reaction.

7.3.2 The Reaction Characteristics and the Optimization of the Reaction Parameters

Figure 7.3a shows the effect of reaction temperature and time on the yield of formate, which is defined as the percentage of formate formed to the initial amount of $NaHCO_3$ based on the carbon content. The yields were obtained from at least triplicate experiments, and the relative error was less than 5%. The production of formate by $NaHCO_3$ proceeded efficiently, and a high formate yield of 40–60% was obtained after only 5 min at above 300°C. After 10 min, the increased rate of formic acid yield decreased significantly. Since almost all the Zn was oxidized to ZnO after only

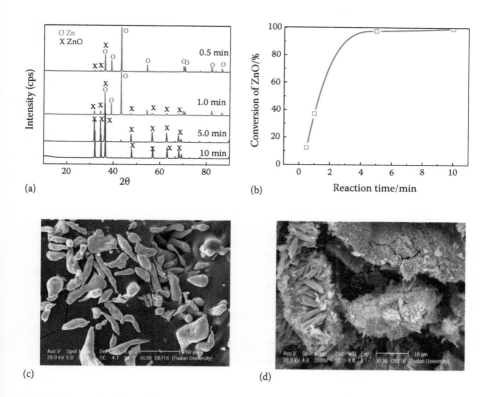

FIGURE 7.2 XRD patterns (a) and the conversion of Zn into ZnO (b) after the reactions for different reaction times. SEM images of Zn powder before the reaction (c) and solid products after the reaction (d) [Zn 10 mmol, NaHCO$_3$ 1 mmol, 300°C, 10 min for (a) and (b), 2 h for (c) and (d)].

10 min, as stated before, it can be concluded that the efficient reduction of NaHCO$_3$ before 10 min should be primarily attributed to the reaction of Zn with H$_2$O. Results under lower reaction temperatures indicated that prolonging the reaction could not increase formate yields efficiently. These results suggest that a temperature greater than 250°C is needed to achieve a high yield of formate.

It has been reported that the initial pH of the solution is an important factor in the production of formate, as pH can affect not only the oxidation of Zn but also the decomposition of the produced formate.[17–19] Thus, the effect of the initial pH of the solution was investigated by adjusting the initial pH with NaOH or HCl. As shown in Figure 7.3b, the initial pH strongly affects the yield of formate. The highest yield of formate was observed at an initial pH of 8.6, which was the natural pH value of aqueous NaHCO$_3$. In the case of a lower, acidic pH of 4.0, the yield of formate decreased to 27.2%. These results indicate that a mildly alkaline pH was favorable for the production of formate.

Figure 7.3c shows the effect of the amount of Zn and NaHCO$_3$ on the yield of formate. An increase in the amount of Zn resulted in a significant increase in formate,

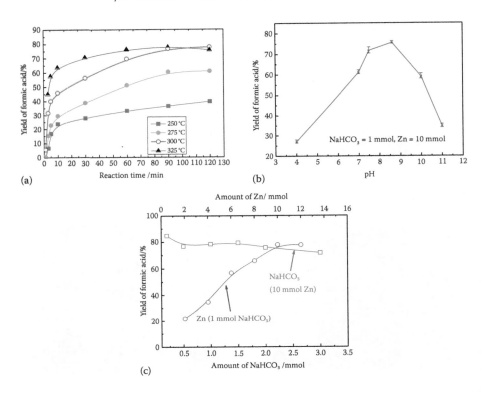

FIGURE 7.3 (a) The effects of temperature and time (NaHCO₃ 1 mmol, Zn 6 mmol). (b) The effect of the initial pH (300°C, 2 h). (c) The effect of the amount of NaHCO₃ and Zn on the yield of formate (300°C, 2 h).

which should be attributed to the promotion of $NaHCO_3$ hydrogenation owing to a higher hydrogen concentration. On the other hand, although a high yield of formate was achieved, a higher Zn/NaHCO₃ ratio of 10:1 was used. Thus, 10 mmol Zn was reacted with various amounts of NaHCO₃ to find the optimal ratio of Zn/NaHCO₃. As shown in Figure 7.3c, the ratio of Zn/NaHCO₃ can be decreased to approximately 3:1 and still produce a formate yield of approximately 70%. The above results clearly showed that formate could be efficiently and selectively produced from NaHCO₃ with Zn under hydrothermal conditions and the reaction parameters affected formate yield quite a lot.

7.4 WATER SPLITTING FOR HYDROGEN PRODUCTION WITH Zn

In the joint system that contains NaHCO₃ and Zn, NaHCO₃ may provide the additional benefit of improving the H₂O and Zn conversions because the oxidation of Zn may be shifted to the right (products) due to the consumption of hydrogen (NaHCO₃ hydrogenation) in the presence of NaHCO₃, as explained in Equation 7.2:

$$H_2O + Zn \rightarrow ZnO + 2H + NaHCO_3 \rightarrow NaCOOH + H_2O \qquad (7.2)$$

TABLE 7.1
Effect of NaHCO$_3$ on the Production of Hydrogen

Entry	NaHCO$_3$/mmol	H$_2$/mL
1	0	52
2[a]	0	70
3	1	74
4	2	78

Note: Reaction conditions: 4 mmol Zn, 300°C, 2 h.

[a] pH of the solution was adjusted to 8.6 with NaOH.

To investigate this topic, the production of hydrogen with and without NaHCO$_3$ was examined. As shown in Table 7.1, hydrogen was formed both in the absence and in the presence of NaHCO$_3$, and NaHCO$_3$ readily promoted the formation of hydrogen from water.

In the oxidation of Zn in water to ZnO, Zn(OH)$_2$ may also form, which is an amphoteric oxide that can be dissolved in alkaline solution to form Zn(OH)$_4^{2-}$. Thus, the increase in hydrogen production in the presence of NaHCO$_3$ could be attributed to the pH increase caused by NaHCO$_3$, which could remove the passivating layer by the dissolution of Zn(OH)$_2$, enabling the rest of Zn to react with water to produce hydrogen. This assumption was supported by a further experiment in which the initial pH value of the solution was adjusted to 8.6 using NaOH in the absence of NaHCO$_3$. The amount of hydrogen collected in this experiment was almost the same as the experiment in the presence of NaHCO$_3$ (Table 7.1, entry 2). Therefore, hydrogen can be effectively produced from the reaction of Zn with water and NaHCO$_3$ is a strong promoter for hydrogen production as it can provide alkaline condition for the promotion of Zn oxidation.

7.5 FORMIC ACID PRODUCTION WITH GASEOUS CO$_2$

After understanding the reaction characteristics and attaining the optimized parameters, the investigations were further conducted with gaseous CO$_2$ to examine whether gaseous CO$_2$ can be used directly. As shown in Table 7.2, almost no formic acid was formed when using gaseous CO$_2$ without the addition of NaOH (Table 7.2, entry 1), which is probably caused by the low dissolution of CO$_2$ in water. Increasing the initial pH of the solution should lead to an increase in the dissolution of CO$_2$ in water. Then, experiments with the addition of NaOH were conducted. The yield of formate, as expected, increased with the increase in the initial pH (Table 7.2, entries 2–5). However, the yield of formate was not high when compared to the yield obtained with NaHCO$_3$. Therefore, to further increase the solubility of CO$_2$ in alkaline solution, the

TABLE 7.2
Yield of Formate with Gaseous CO_2 under Various Reaction Conditions

	Reaction Conditions				
Entry	CO_2/mmol	Initial pH	NaOH	Zn/mmol	Yield/%
1	2	6.7	Without	6	0.02
2	2	8.8	With	6	0.03
3	2	11.4	With	6	0.74
4	2	13.5	With	6	5.9
5	2	14	With	6	16.6

Note: Reaction conditions: 300°C, 2 h.

alkaline solution (pH 14) was maintained at room temperature after the introduction of CO_2 before the hydrothermal reaction. As a result, the yield of formate increased with the dissolution time. In the meantime, the pH of the solution decreased incessantly, and the highest formate yield was obtained when the pH value of the solution changed to 9.2, which was very close to the pH value of the $NaHCO_3$ solution (Figure 7.4). These results indicate that gaseous CO_2 can be directly used to produce formate when dissolved in alkaline solution ahead of the reaction, and the formation of formate occurred mainly from the hydrogenation of HCO_3^-. The calculated equilibrium distribution of CO_2, HCO_3^-, CO_3^{2-}, and $HCOO^-$ at 300°C, 350 bars reported that HCO_3^- and $HCOO^-$ were the predominantly dissolved carbon species at a moderately alkaline pH,[17,19] which supported the above assumption arguably.

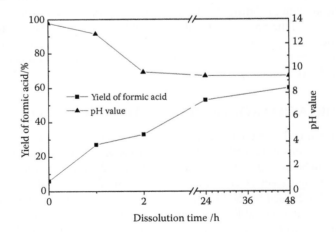

FIGURE 7.4 Yield of formic acid with the dissolution of gaseous CO_2 (gaseous CO_2 1 mmol, NaOH 1 mmol, Zn 6 mmol, 300°C, 2 h, dissolution of gaseous CO_2 was performed at room temperature).

7.6 PROPOSED MECHANISM FOR FORMATE PRODUCTION

In the reaction of converting $NaHCO_3$ into formate with Zn under hydrothermal conditions, no catalyst was added. However, it is generally known that a catalyst is needed in the hydrogenation of CO_2. One possible explanation for these observations is that ZnO, formed by the oxidation of Zn under hydrothermal conditions, plays an autocatalytic role in the reaction, as ZnO is traditionally a good hydrogenation catalyst.[20,21] To assess whether the formed ZnO acts as an autocatalyst, Zn was substituted with gaseous hydrogen and ZnO. Reagent-grade ZnO, as well as both dry and wet ZnO collected after the Zn reactions were completed, were used in these experiments. A two-step reaction was investigated for the experiments with wet ZnO. In the first step, Zn reacted with water, and H_2 and $NaHCO_3$ were added to the solution in the second step. Compared to the formate yields obtained without ZnO, the yield of formate with ZnO, particularly for wet ZnO collected after the Zn reaction, was slightly higher (Table 7.3), indicating that the added ZnO, even for the wet ZnO collected after the Zn reaction, only provides a very weak catalytic activity. These results suggest that when Zn reacted under hydrothermal conditions, some active intermediate may be formed for the hydrogenation reaction of $NaHCO_3$. As already discussed in Figures 7.2a and 7.3a, Zn was completely oxidized to ZnO after 10 min and the formate production proceeded rapidly in the first 10 min and then slowed down thereafter. These results together with the formate production with hydrogen and ZnO indicated that the in situ formed ZnO probably catalyzed the slow formation of formate at reaction time over 10 min; however, in the first 10 min, another reaction mechanism, which contains the active hydrogenation intermediate, should dominate the rapid formate generation.

TABLE 7.3
Formate Yield with $NaHCO_3$ in the Presence and Absence ZnO

Run	Reductant	Reductant/$NaHCO_3$ (mmol)	Additives	Yield/%
1	Zn	6/1	Without	57.5
2	H_2	6/1	Without	13.0
3	H_2	6/1	Reagent ZnO[a]	16.0
4	H_2	6/1	Collected dry ZnO[b]	15.5
5	H_2	6/1	Collected wet ZnO[c]	23.0

Note: The amount of ZnO in mole is the same as the reductant for all experiments.

[a] Reagent ZnO: In powder with 200-mesh size.

[b] The solid residue collected after the reaction with Zn, which was washed with deionized water several times, filtrated, and dried in air.

[c] A two-step reaction was used; in the first step, Zn reacted with H_2O; in the second step, $NaHCO_3$ and H_2 were added to the reactor.

Infrared (IR) absorption peaks of solid samples after the reactions at different reaction times in the presence and absence of $NaHCO_3$ were examined to figure out the possible hydrogenation intermediate. As shown in Figure 7.5, no significant IR absorption peaks appeared for bulk ZnO. When Zn alone was reacted with water without the addition of $NaHCO_3$, two absorption peaks at 3336.8 and 1640 cm^{-1} were observed in the solid residue, which should be attributed to the stretching vibration of the O–H complex and Zn–H complex in H–Zn⋯O–H, respectively,[22] suggesting that an active intermediate structure of H–Zn⋯O–H is probably formed because of the water-splitting reaction on Zn. It has been reported that Zn hydride species (Zn–H) can be produced in the reaction of Zn with H_2O at 700 K.[23] Zn hydride species (Zn–H) can act as a possible source of active hydrogen for hydrogenation reactions, such as the Cu/ZnO-catalyzed synthesis of methanol from a mixture of CO, CO_2, and H_2[24] and the formic acid production from CO_2.[25,26] Therefore, the hydrogen in

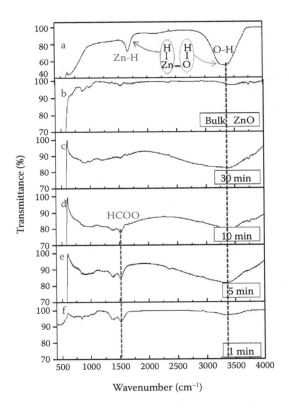

FIGURE 7.5 FT-IR absorption spectra of (a) wet solid sample after reactions of Zn with water in the absence of $NaHCO_3$ (Zn 10 mmol, 300 °C, 10 min), (b) bulk ZnO, and (c through f) solid samples after reactions of Zn with water in the presence of $NaHCO_3$ ($NaHCO_3$ 1 mmol, Zn 10 mmol, 300°C).

the Zn–H complex could act as an active hydrogen source for the hydrogenation of $NaHCO_3$ to formate owing to the weak Zn–H bond.

When $NaHCO_3$ was added to the reaction, the IR absorption peak for O–H complex in H–Zn···O–H was still present; however, the peak for the Zn–H complex disappeared. Interestingly, the two peaks at 1390 and 1510 cm^{-1} corresponded respectively to the asymmetrical and symmetrical stretching of the absorbed $HCOO^-$ observed.[27] These results indicate that the absence of the Zn–H complex absorption peak is probably because the hydrogen in Zn–H reacted with $NaHCO_3$ and resulted in the formation of formate.

Based on the obtained results, a possible reaction mechanism of the Zn autocatalytic HCO_3^- reduction into formate in high-temperature water is proposed, as shown in Figure 7.6. First, H–Zn···O–H complex **1** is generated by the oxidation of Zn in water (reaction **I**). The subsequent nucleophilic attack of the anionic proton of the formed complex **1** to the bicarbonate ion **2** yields the compound formate (complex **4**) along with Zn complex **5**, and this complex then loses water to form ZnO **6** (an SN2-like mechanism).

Moreover, when Zn is oxidized to ZnO, the mechanism of hydrogen absorption onto ZnO (reaction **II**) may also occur because a small amount of formate can be formed by using H_2 + ZnO + $NaHCO_3$ (Table 7.3). A similar structure of H–Zn···O–H (complex **7**) is probably formed as a result of the chemisorption of hydrogen on ZnO. However, the hydrogen absorption efficiency may not be high, as only approximately 5–10% surface sites of ZnO can be covered by hydrogen, which may explain why hydrogen cannot reduce $NaHCO_3$ as efficiently as Zn. As reaction **II** is a diffusion limited reaction, it is possible to be favored by prolonging the reaction time. As a result, the rapid formation of formate in the first 10 min is dominated by reaction **I**, which is the predominant reaction route for the hydrogenation of $NaHCO_3$, and the slow formate generation after 10 min is mainly contributed by reaction **II**, which produces minor formate.

7.7 QUANTUM CHEMICAL CALCULATIONS OF THE REACTION MECHANISM

Further, density functional theory (DFT) calculations on the crucial intermediate (Zn–H) and the mechanism of the formate formation were performed. A model system consisting of a Zn_5 cluster with two H_2O molecules was adopted for all theoretical investigations. The Zn_5 model contains all the three site types, that is, step (linear), terrace (surface), and kink (vertex), which resulted in a total of five possible positions of H_2O fragments on the Zn_5 model (Figure 7.7). All the possible conformations of the Zn_5 cluster were first optimized with all the possible fragments (H^+, OH^-, and H_2O) generated through the fragmentation of two H_2O molecules, which contains a total of 859 conformations (Table 7.4). Adopting the free energy as the screening criteria, relative to the reference consisting of the optimized Zn_5 cluster and the two non-fragmented H_2O molecules, calculations showed the existence of the energetically favorite Zn–H intermediate at 573 K as a product of the Zn_5 + $2H_2O$ reactions.

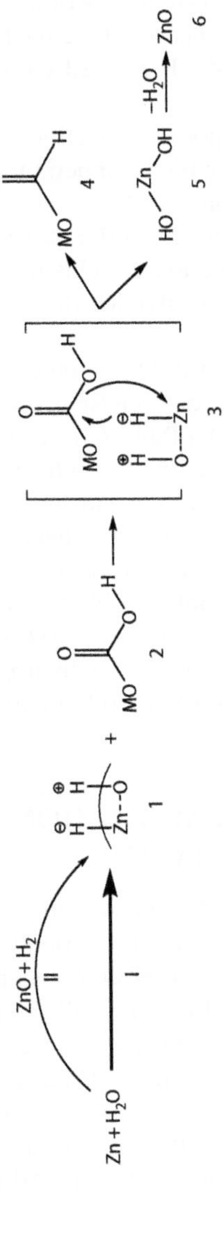

FIGURE 7.6 Proposed mechanism of reduction of HCO_3^- into formate with Zn (M represents the metal cation in bicarbonate).

Hydrothermal CO₂ Reduction with Zinc to Produce Formic Acid

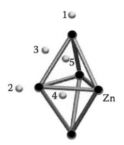

FIGURE 7.7 Five possible positions of water fragments (H_2O, H^+, OH^-, and O^{2-}) on Zn_5 cluster (water fragments are depicted with yellow balls).

TABLE 7.4
Water Fragments Distribution Modes on Zn5 Cluster

Mode	Structures	Amounts
1	Zn_5, {H_2O, H_2O}	20
2	Zn_5, {H_2O, H^+, OH^-}	40
3	Zn_5, {H_2O, H^+, H^+, O^{2-}}	228
4	Zn_5, {H^+, OH^-, H^+, OH^-}	29
5	Zn_5, {H^+, OH^-, H^+, H^+, O^{2-}}	179
6	Zn_5, {H^+, H^+, O^{2-}, H^+, H^+, O^{2-}}	363
Total		859

The feasibility of the formate formation through the Zn–H intermediate was determined by the calculation of the geometry of the transition state (TS) and the energy for the formate formation. Based on the calculated free energy at 573 K for the Zn–H formation, Pattern 4 in Figure 7.8 was adopted as the most stable form of Zn–H and used in the TS search. As shown in Figure 7.9, the activation energy of the TS from the initial state (Zn–H and HCO_3^-) is 24.1 kcal/mol, which can be readily achieved under 573 K. The geometry and the Mulliken charge on the TS reflect the production of $HCOO^-$ from HCO_3^-. The Zn–H bond distance is approximately 1.94 Å. The charge of H in the Zn–H species is −0.221. This charge is assigned to be the hydride rather than the proton. An important implication of the TS geometry and charge distribution is that this is an SN2-like reaction. As the hydride of Zn–H approaches the carbon of HCO_3^-, the OH^- separates from the carbon atom. Thus, as $HCOO^-$ is formed, the OH^- is drawn to the Zn_5 cluster, as shown in Figure 7.6.

The results of intrinsic reaction coordinate (IRC) calculations and the shapes of the HOMO and LUMO of the TS are shown in Figure 7.10. As shown in Figure 7.10a, the plain IRC curve of the TS led to a smooth transformation of the reactant Zn–H + HCO_3^- to the product formate. The reaction energy barrier of HOMO to LUMO for the formation of formate from HCO_3^- was not very high, which indicates that formate could be easily produced once the Zn–H species were produced. In Figure 7.10b, the occupied HOMO of the TS shows a bonding interaction between the carbon (C) atom of HCO_3^- and the hydrogen (H) atom of the Zn–H intermediate,

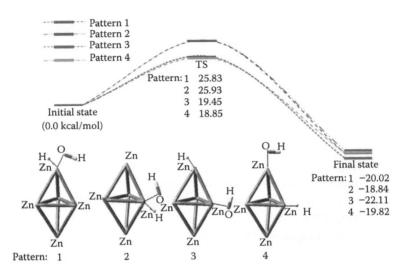

FIGURE 7.8 Calculated free energy at 573 K for the Zn–H formation in four patterns.

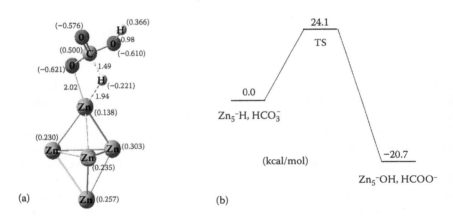

FIGURE 7.9 The geometries of TS (a) and activation energy of Zn–H + $HCO_3^- \rightarrow Zn_5$–OH + $HCOO^-$ (b).

whereas the unoccupied LUMO of the TS shows an anti-bonding character between the carbon of HCO_3^- and the OH^- (there is also an anti-bonding character in the LUMO of the Zn–H bond). The HOMO and LUMO of the TS clearly demonstrated that formate production from HCO_3^- followed an SN2-like mechanism. Based on these theoretical calculations, it could be concluded that the formation of the Zn–H complex was energetically favorable, and the TS of the reaction could be achieved readily under 573 K, which could lead smoothly to a formate production via an SN2-like mechanism.

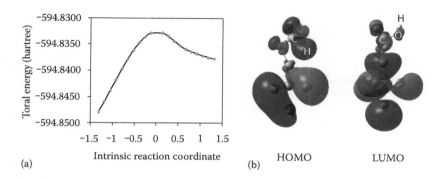

FIGURE 7.10 IRC calculation (a) and HOMO and LUMO orbital shapes (b) of the TS.

7.8 Zn–ZnO CYCLE AND ASSESSMENT OF THE ENERGY CONVERSION EFFICIENCY

It has been reported that ZnO can be readily reduced into Zn using concentrated solar energy,[8,11] which is a promising technology for industrial utilization in the near future.[8] Thus, the ZnO formed in the reduction of CO_2 with water splitting with Zn as a reductant can be reduced back into Zn (0) by solar energy; that is, by combining the solar reduction of ZnO to Zn with the presented method of Zn reduction of HCO_3^- to formate, a highly efficient method for converting the solar energy to chemical fuels could be achieved (Figure 7.1).

The energy conversion efficiency from solar to fuel (formic acid) was then evaluated. The chemical equation of formic acid production from CO_2 Zn and water can be written as Equation 7.1. On the other hand, the equation of solar reduction of ZnO to Zn can be written as

$$ZnO + \text{solar energy} \rightarrow Zn \tag{7.3}$$

Therefore, the solar-to-formic acid energy conversion efficiency can be expressed as

$$\eta_{\text{solar-HCOOH}} = \eta_1 \times \eta_2, \tag{7.4}$$

where η_1 is the energy conversion efficiency of Equation 7.1 and η_2 is the energy conversion efficiency of Equation 7.3. Since Equation 7.1 represents the chemical energy of Zn transformed into the chemical energy of formic acid, η_1 can be expressed as

$$\eta_1 = \frac{M_{\text{HCOOH}} \times \Delta H_{\text{HCOOH}}}{M_{\text{Zn}} \times \Delta H_{\text{Zn}}}, \tag{7.5}$$

where M_{HCOOH} is the molar production of formic acid, M_{Zn} is the molar consumption of Zn, and ΔH_{HCOOH} and ΔH_{Zn} are the higher heating values of formic acid and

Zn, respectively. As discussed in Section 7.3, more than 70% conversion efficiency from CO_2 to formic acid can be achieved under the reaction condition of $NaHCO_3$/Zn = 1:3, which means, that 3 mol Zn could yield 0.7 mol formic acid. Thus, by substituting the higher heating values of formic acid and ZnO, the energy conversion efficiency of Equation 7.1 can be estimated as

$$\eta_1 = \frac{0.7 \times (-254.34)}{3 \times (-350.46)} \times 100\% = 16.9\%. \quad (7.6)$$

An energy conversion efficiency of 29–36% for $ZnO-Zn-H_2$ by solar energy can be achieved according to the literature.[8] If the lower energy conversion efficiency of 30% for $ZnO-Zn-H_2$ by solar energy was used as the η_2, the overall solar-to-formic acid energy conversion efficiency can be obtained at around 5%. Thus, the integrated technology from coupling the proposed hydrothermal production of formic acid and the reduction of ZnO into Zn by solar energy can be compared with related technologies in the literature.[28,29]

7.9 CONCLUSIONS

In summary, the recent advances in reduction of CO_2 into formic acid by water dissociation with metallic Zn are reviewed. These results show that $NaHCO_3$ or CO_2 can be reduced efficiently and rapidly into formic acid by water splitting with metallic Zn under hydrothermal conditions, and a high yield (70–80%) of formate from $NaHCO_3$ with nearly 100% selectivity can be achieved. Based on the experimental results and DFT studies, the in situ formed intermediate of H–Zn⋯O–H during the water dissociation with Zn played a key role in converting $NaHCO_3$ into formate. By coupling the proposed method with the solar-driven reduction of ZnO to Zn, a Zn–ZnO cycle for the continuous reduction of CO_2 into formic acid with water-derived hydrogen can be achieved; that is, a highly efficient solar–fuel process can be expected.

ACKNOWLEDGMENTS

The authors thank the financial support of the National Natural Science Foundation of China (No. 21277091), the State Key Program of National Natural Science Foundation of China (No. 21436007), Key Basic Research Projects of Science and Technology Commission of Shanghai (No. 14JC1403100), and China Postdoctoral Science Foundation (No. 2013 M541520).

REFERENCES

1. Jessop, P. G.; Joó, F.; Tai, C.-C. 2004. Recent advances in the homogeneous hydrogenation of carbon dioxide. *Coordination Chemistry Reviews* 248 (21–24): 2425–2442.
2. Sakakura, T.; Choi, J. C.; Yasuda, H. 2007. Transformation of carbon dioxide. *Chemistry Reviews* 107 (6): 2365–2387.

3. Wang, W.; Wang, S.; Ma, X.; Gong, J. 2011. Recent advances in catalytic hydrogenation of carbon dioxide. *Chemical Society Reviews* 40 (7): 3703–3727.
4. Yadav, R. K.; Baeg, J. O.; Oh, G. H.; Park, N. J.; Kong, K. J.; Kim, J.; Hwang, D. W.; Biswas, S. K. 2012. A photocatalyst-enzyme coupled artificial photosynthesis system for solar energy in production of formic acid from CO_2. *Journal of the American Chemical Society* 134 (28): 11455–11461.
5. Steinfeld, A. 2005. Solar thermochemical production of hydrogen—A review. *Solar Energy* 78 (5): 603–615.
6. Kodama, T.; Gokon, N. 2007. Thermochemical cycles for high-temperature solar hydrogen production. *Chemistry Reviews* 107 (10): 4048–4077.
7. Gálvez, M. E.; Loutzenhiser, P. G.; Hischier, I.; Steinfeld, A. 2008. CO_2 splitting via two-step solar thermochemical cycles with Zn/ZnO and FeO/Fe_3O_4 redox reactions: Thermodynamic analysis. *Energy & Fuels* 22 (5): 3544–3550.
8. Steinfeld, A. 2002. Solar hydrogen production via a two-step water-splitting thermochemical cycle based on Zn/ZnO redox reactions. *International Journal of Hydrogen Energy* 27 (6): 611–619.
9. Steinfeld, A. 1998. Solar-processed metals as clean energy carriers and water-splitters. *International Journal of Hydrogen Energy* 23 (9): 767–774.
10. Haueter, P.; Moeller, S.; Palumbo, R.; Steinfeld, A. 1999. The production of zinc by thermal dissociation of zinc oxide—Solar chemical reactor design. *Solar Energy* 67 (1–3): 161–167.
11. Chueh, W. C.; Falter, C.; Abbott, M.; Scipio, D.; Furler, P.; Haile, S. M.; Steinfeld, A. 2010. High-flux solar-driven thermochemical dissociation of CO_2 and H_2O using nonstoichiometric ceria. *Science* 330 (6012): 1797–1801.
12. Chueh, W. C.; Haile, S. M. 2009. Ceria as a thermochemical reaction medium for selectively generating syngas or methane from H_2O and CO_2. *ChemSusChem* 2 (8): 735–739.
13. McCammon, C. 2005. The paradox of mantle redox. *Science* 308 (5723): 807–808.
14. Holm, N. G.; Charlou, J. L. 2001. Initial indications of abiotic formation of hydrocarbons in the Rainbow ultramafic hydrothermal system, Mid-Atlantic ridge. *Earth and Planetary Science Letters* 191 (1–2): 1–8.
15. McCollom, T. M.; Seewald, J. S. 2001. A reassessment of the potential for reduction of dissolved CO_2 to hydrocarbons during serpentinization of olivine. *Geochimica et Cosmochimica Acta* 65 (21): 3769–3778.
16. Steinfeld, A.; Brack, M.; Meier, A.; Weidenkaff, A.; Wuillemin, D. 1998. A solar chemical reactor for co-production of zinc and synthesis gas. *Energy* 23 (10): 803–814.
17. Jin, F.; Yun, J.; Li, G.; Kishita, A.; Tohji, K.; Enomoto, H. 2008. Hydrothermal conversion of carbohydrate biomass into formic acid at mild temperatures. *Green Chemistry* 10 (6): 612–615.
18. Yasaka, Y.; Yoshida, K.; Wakai, C.; Matubayasi, N.; Nakahara, M. 2006. Kinetic and equilibrium study on formic acid decomposition in relation to the water-gas-shift reaction. *Journal of Physical Chemistry A* 110 (38): 11082–11090.
19. Peterson, A. A.; Vogel, F.; Lachance, R. P.; Fröling, M.; Antal, J. M. J.; Tester, J. W. 2008. Thermochemical biofuel production in hydrothermal media: A review of sub- and supercritical water technologies. *Energy & Environmental Science* 1 (1): 32–65.
20. Arena, F.; Barbera, K.; Italiano, G.; Bonura, G.; Spadaro, L.; Frusteri, F. 2007. Synthesis, characterization and activity pattern of $Cu–ZnO/ZrO_2$ catalysts in the hydrogenation of carbon dioxide to methanol. *Journal of Catalysis* 249 (2): 185–194.
21. Liang, X.-L.; Dong, X.; Lin, G.-D.; Zhang, H.-B. 2009. Carbon nanotube-supported Pd–ZnO catalyst for hydrogenation of CO_2 to methanol. *Applied Catalysis B: Environmental* 88 (3–4): 315–322.

22. Chauvin, C.; Saussey, J.; Lavalley, J. C.; Djega-Mariadassou, G. 1986. Definition of polycrystalline ZnO catalytic sites and their role in CO hydrogenation. *Applied Catalysis* 25 (1–2): 59–68.
23. Umemoto, H.; Tsunashima, S.; Ikeda, H.; Takano, K.; Kuwahara, K.; Sato, K.; Yokoyama, K.; Misaizu, F.; Fuke, K. 1994. Nascent internal state distributions of ZnH(X2Σ+) produced in the reactions of Zn(41P1) with some alkane hydrocarbons. *The Journal of Chemical Physics* 101 (6): 4803–4808.
24. Zapol, P.; Jaffe, J. B.; Hess, A. C. 1999. Ab initio study of hydrogen adsorption on the ZnO surface. *Surface Science* 422 (1–3): 1–7.
25. Sattler, W.; Parkin, G. 2012. Zinc catalysts for on-demand hydrogen generation and carbon dioxide functionalization. *Journal of the American Chemical Society* 134 (42): 17462–17465.
26. Yu, K. M.; Yeung, C. M.; Tsang, S. C. 2007. Carbon dioxide fixation into chemicals (methyl formate) at high yields by surface coupling over a Pd/Cu/ZnO nanocatalyst. *Journal of the American Chemical Society* 129 (20): 6360–6361.
27. Tardio, J.; Bhargava, S.; Prasad, J.; Akolekar, D. B. 2005. Catalytic wet oxidation of the sodium salts of citric, lactic, malic and tartaric acids in highly alkaline, high ionic strength solution. *Topics in Catalysis* 33 (1–4): 193–199.
28. Barton, E. E.; Rampulla, D. M.; Bocarsly, A. B. 2008. Selective solar-driven reduction of CO_2 to methanol using a catalyzed p-GaP based photoelectrochemical cell. *Journal of the American Chemical Society* 130 (20): 6342–6344.
29. Varghese, O. K.; Paulose, M.; Latempa, T. J.; Grimes, C. A. 2009. High-rate solar photocatalytic conversion of CO_2 and water vapor to hydrocarbon fuels. *Nano Letters* 9 (2): 731–737.

8 Autocatalytic Hydrothermal CO$_2$ Reduction with Manganese to Produce Formic Acid

Lingyun Lyu, Fangming Jin, and Guodong Yao

CONTENTS

8.1 Introduction 109
8.2 Materials and Methods 111
    8.2.1 Materials 111
    8.2.2 Experimental Procedure 111
    8.2.3 Product Analysis 111
8.3 Results and Discussion 112
    8.3.1 Potential of CO$_2$ Reduction with Mn 112
    8.3.2 Characteristics of the Reaction of Dissociation of Water for CO$_2$ Reduction with Mn and the Parameter Design of the High Yield of Formic Acid 115
    8.3.3 CO$_2$ Role in Improving Hydrogen Production from Water 118
    8.3.4 Possible Mechanism of Mn Oxidation in Water and the Role of Mn$_x$O$_y$ in the Formation of Formic Acid 119
    8.3.5 Investigation of Formic Acid Formation via HCO$_3^-$ or Gaseous CO$_2$.... 121
8.4 Conclusions and Prospects 124
References 125

8.1 INTRODUCTION

Atmospheric CO$_2$ is rising at a fast rate; the average annual increasing rate of atmospheric CO$_2$ level is 2.11 parts per million (ppm) per year in 2005–2014, which is more than double compared with that in the 1960s (1.06 ppm per year) according to the Intergovernmental Panel Climate Change's report (IPCC 2015) (Cox et al. 2000; Park and Fan 2004). The "Greenhouse effect" phenomenon can be attributed to CO$_2$ emissions from fossil fuel consumption of the human society to meet the global energy demands of economic growth (Kone and Buke 2010; Meng and Niu

2011). Presently, the concentration of CO_2 is increasing at an unprecedented rate, and many sectors, such as agriculture, food production, industry, tourism, and health, are affected by this important phenomenon (Grimm et al. 2013; Patz et al. 2005; Sumaila et al. 2011; Wheeler and Braun 2013). Furthermore, the melting of glaciers, the rising of the ocean water levels and vaporization in freshwater resources as heat rises harm the natural balance and threaten the ecological environment.

A great deal of development effort has already been expended in the areas of reduction of CO_2 concentration in the atmosphere, among which solar technology is the ideal solution (Jessop et al. 2004; Wang et al. 2011; Yadav et al. 2012). Artificial photosynthesis, in which solar energy is converted into chemical energy for renewable, nonpolluting fuels and chemicals, is regarded as one of the most promising methods. However, challenges remain in the direct conversion of solar energy into chemical energy, such as poor conversion efficiencies and poor product selectivity. Developing an efficient solar-to-fuel conversion process is a great and fascinating challenge (Indrakanti et al. 2009; Roy et al. 2010; Takeda et al. 2008; Varghese et al. 2009). In contrast to direct solar-to-fuel conversion, an integrated system should be expected to improve the efficiency of artificial solar-to-fuel conversion. Recently, some interesting integrated technologies, such as a solar two-step water-splitting thermochemical cycle based on redox of metals/metal oxides, have been reported, in which Fe/Fe_3O_4, Zn/ZnO, $Mn(III)/Mn(II)$, and $Mn(IV)/Mn(II)$ and even Mg_xO_y/Mg have been achieved using solar energy. Thus, water splitting for hydrogen production was significantly higher than direct use of solar energy, which would be one of most promising approaches to increase artificial photosynthetic efficiency (Galvez et al. 2008; Haueter et al. 1999; Hu 2012; Uchida 2012).

Previous research has shown that water splitting for the reduction of CO_2 to formic acid with Fe could be conducted. However, even with the addition of a nickel catalyst, the highest formic acid yield on a carbon basis was approximately 16% (Jin et al. 2011; Tian et al. 2007; Wu et al. 2009). Mn, as a first-row transition metal, has an extraordinarily appealing coordination chemistry because of its reactive redox nature (Cornia et al. 1999). As a key element in photosynthesis, Mn can mediate the splitting of water to provide the necessary electrons for photosynthesis (Neelameggham 2008, 2009; Shabala 2009). In addition, Mn plays a significant role in the synthesis of catalysts for CO_2 hydrogenation and the Fischer–Tropsch process (Sheshko and Serov 2012). Moreover, many researchers have reported a redox process for ZnO/Zn, Mn(III)/Mn(II), and Mn(IV)/Mn(II) reaction using solar energy (Galvez et al. 2008; Haueter et al. 1999). Recently, Uchida have also demonstrated that MgO/Mg can be circulated by solar power concentration using laser technology (Uchida 2012). The redox of MgO/Mg suggests that the reduction of Mn_xO_y should be much easier than that of MgO by using solar energy because Mg is more active that Mn. Thus, circulation of Mn/Mn_xO_y should be achieved. Therefore, metallic Mn may have a much more significant implication than Fe in water splitting for the conversion of CO_2.

To the best of our knowledge, no study has reported the utilization of metallic Mn as an efficient reductant reacting with water to produce hydrogen for the conversion of CO_2. With the above concept of highly efficient dissociation of water based on redox of metals/metal oxides for CO_2 reduction with a high yield, using Mn as reductant to produce hydrogen for CO_2 reduction was investigated because the metal oxide/metal cycle with solar energy has been well studied currently. The results showed that Mn performed very well in the CO_2 conversion process. These new findings are reported in this chapter.

8.2 MATERIALS AND METHODS

8.2.1 Materials

Mn powder was purchased from Aladdin Chemical Reagent, and $NaHCO_3$ was purchased from Sinopharm Chemical Reagent Co., Ltd. In this study, $NaHCO_3$ was used as a CO_2 source to simplify handling. Gaseous CO_2 and H_2 (>99.995%) were obtained from Shanghai Poly-Gas Technology Co., Ltd. Deionized water was used throughout the study.

8.2.2 Experimental Procedure

All of the experiments were conducted in a series of batch stainless steel (SUS-316) tubing reactors (9.525 mm [3/8 in.] o.d., 1-mm wall thickness, 120 mm long] with end fittings, providing an inner volume of 5.7 mL. Teflon-lined reactors were used to examine the effect of reactor wall on the reaction for CO_2 reduction. The schematic drawing can be found elsewhere (Jin et al. 2001, 2005). The experimental procedure was as follows. The desired amounts of Mn, $NaHCO_3$, and deionized water were added to the reactor chamber. The reactor tube was sealed and then immersed into a salt bath that had been preheated to the desired temperature (the reactor was kept shaking during the reaction). After the preset reaction time, the reactor tube was then taken out from the salt bath and put into a cold water bath near the salt bath to quench the reaction immediately. Finally, after cooling the reactor to room temperature, the reaction mixture (the gas, liquid, and solid samples) was collected and the liquid sample was filtered through a 0.22-μm syringe for analysis. Water filling was defined as the ratio of the volume of the water put into the reactor to the inner volume of the reactor, and reaction time was defined as the duration of time the reactor was kept in the salt bath.

8.2.3 Product Analysis

The yield of formic acid was defined as the percentage of formic acid and the initial $NaHCO_3$ on a carbon basis as follows:

$$Y = \frac{C_F}{C_S}, \quad (8.1)$$

where C_F and C_S are the amount of carbon in formic acid and in the initial $NaHCO_3$ added to the reactor. Liquid samples were analyzed by HPLC (high-performance liquid chromatography), TOC (total organic carbon), and GC/MS (gas chromatography/mass spectroscopy). HPLC analysis was performed on KC-811 columns (SHODEX) with an Agilent Technologies 1200 system, which was equipped with a tunable UV/

Vis absorbance detector adjusted to 210 nm and a differential refractometer detector, and the system used a 2 mmol/L $HClO_4$ solution as the mobile phase at a flow rate of 1.0 mL/min. TOC was analyzed using a Shimadzu TOC 5000A. Gas samples were analyzed by GC/TCD (gas chromatography/thermal conductivity detector). The Agilent 7890 GC/MS system, which was equipped with a 5985C inert MSD and a triple-axis detector, was used to investigate other possible chemicals in liquid samples. The solid samples were washed with deionized water three times to remove impurities and with ethanol three times to make the solid sample dry quickly. The samples were then naturally dried at room temperature and characterized using XRD (x-ray diffraction). XRD analyses were performed on a Bruker D8 Advance x-ray diffractometer. The step scan covered angles of 10–80° (2 θ) at a rate of 2°/s.

8.3 RESULTS AND DISCUSSION

8.3.1 Potential of CO_2 Reduction with Mn

In our previous study, the dissociation water for the reduction of CO_2 to formic acid has been successfully conducted when Fe was used (Wu et al. 2009). In this study, to test whether CO_2 could be reduced to useful organics and to determine the reduction products when Mn was used as reagent, experiments with $NaHCO_3$ and Mn powder (200 mesh) were conducted at 300°C for 2 h with a water filling of 35%, which is an ideal condition for obtaining a high yield of formic acid when using Fe as a reductant. Liquid samples were analyzed by HPLC, GC/MS, and TOC. As shown in Figure 8.1, HPLC analysis showed that the main product was formic acid, and GC/MS analysis validated this result, with very little acetic acid detected. Additionally, the organic carbon in the liquid samples was detected by TOC analysis. The results showed that the selectivity of the production of formic acid was more than 98%, which was defined as the percentage of the amount of carbon in the formic acid comparable to the total carbon in the liquid sample. Analysis of gas samples by GC/TCD showed that no organic product was produced, and only hydrogen and a small amount of CO_2 were detected. These results indicated that formic acid was the main product from CO_2 in the presence of Mn. Quantitative analysis of the products obtained under this condition showed that the yield of formic acid was 43%. Comparing with the reaction that used Fe as a reductant (the highest formic acid yield was only 16% with nickel as a catalyst) (Jin et al. 2011; Wu et al. 2009), the yield of formic acid with Mn was much higher, indicating that Mn was more efficient in reducing CO_2 into formic acid than Fe. Subsequently, experiments with different sizes of Mn powder were first conducted, and the results demonstrated that the yield of formic acid has no evident change with the different sizes of Mn powder ranging from 50 to 1400 mesh, as shown in Figure 8.2. The 200-mesh Mn powder was chosen in this study. Moreover, to examine the effects of the reactor wall of SUS316, experiments conducted in a series of Teflon-lined reactor that has no metal in the reactor wall were conducted. Considering that the reaction temperature and time are very sensitive to the yield of formic acid from CO_2, that the temperature limitation of the Teflon material is no more than 250°C, and that the Teflon reactor uses an oven as a heater, which leads to a slow heating rate, all of the experiments in the Teflon-lined reactor

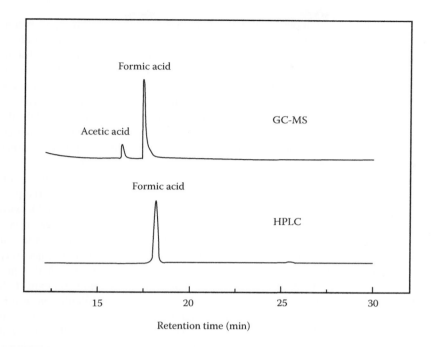

FIGURE 8.1 HPLC and GC/MS chromatogram of liquid sample after reaction (temperature: 300°C; time: 2 h; NaHCO$_3$: 1 mmol; Mn: 8 mmol; water filling: 35%).

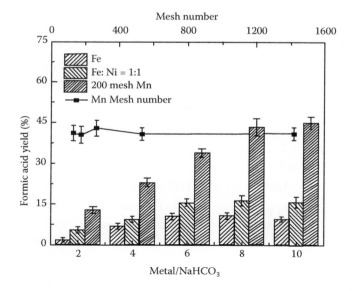

FIGURE 8.2 Effect of the amount of metal reductant ratio and Mn size on the formic acid yield (temperature: 300°C; time: 2 h; NaHCO$_3$: 1 mmol; water filling: 35%).

were conducted at a low temperature of 250°C for a long period of 12 h. As shown in Table 8.1, the yield of formic acid was considerably high when using metallic manganese (Mn) without catalyst. However, when using metallic iron (Fe) and magnesium (Mg), without the addition of catalyst Ni, the formic acid yields were only 1.8% for Fe and 2% for Mg; even for Mg, in which a large amount of hydrogen was produced, the yield of formic acid increased evidently when the catalyst Ni was added. Hence, the results showed that no significant catalytic role of the reactor wall of SUS 316 was observed, and a high yield of formic acid from CO_2 with Mn as catalyst resulted from the catalytic role of some components formed in the reaction process, rather than the role of the reactor wall of SUS 316.

In addition, the Gibbs free energy and enthalpy value assessment for conversion of CO_2 into formic acid with metallic Mn was also examined. According to the proposed mechanism, the overall reaction of the hydrothermal reduction of CO_2 with Mn can be written as Equation 8.2, and the calculated reaction heat and free energy using available thermodynamic data are negative. The calculation results indicated that the dissociation of water for CO_2 reduction using Mn as reductant is not only spontaneous but also exothermic. It is known that for an exothermic reaction, its equilibrium constant (K_{eq}) will decrease with an increase in temperature. As expected, the calculated $K_{eq\ (600\ K)}$ is significantly lower than $K_{eq\ (298\ K)}$. The calculation results conformed to the rule of exothermic reaction, that is, that K_{eq} will decrease with an increase in temperature because the reverse (endothermic) reaction will be favored with the addition of heat. ΔG will be less negative with an increase in temperature, which will cause K_{eq} to decrease.

$$Mn + CO_2 + H_2O \rightarrow MnO + HCOOH \tag{8.2}$$

$\Delta G^\circ_{298} = -23.01$ kJ/mol $\qquad \Delta H^\circ_{298} = -114.75$ kJ/mol

$\Delta G^\circ_{600} = -10.78$ kJ/mol $\qquad \Delta H^\circ_{600} = -115.78$ kJ/mol

$K_{eq\ (298K)} = e^{9.29} = 10,829 \qquad K_{eq\ (600K)} = e^{2.16} = 8.67$

TABLE 8.1
Formic Acid Yields with Teflon-Lined Reactor

Run	Y_{HCOOH} (%)	Reductant	$NaHCO_3$	Ni
1	17.9	24 mmol Mn	4 mmol	Without
2	1.8	24 mmol Fe	4 mmol	Without
3	2.0	16 mmol Mg	4 mmol	Without
4	8.0	24 mmol Fe	4 mmol	With
5	12.5	16 mmol Mg	4 mmol	With

Note: Reaction condition: 250°C, 12 h, water filling 35%.

8.3.2 Characteristics of the Reaction of Dissociation of Water for CO_2 Reduction with Mn and the Parameter Design of the High Yield of Formic Acid

The effects of the reaction conditions, such as initial amounts of Mn and $NaHCO_3$, reaction temperature, reaction time, and water filling, were studied to first investigate the characteristics of the dissociation of water for CO_2 reduction. As shown in Figure 8.3a, the initial amount of Mn strongly affected the formic acid yield at a fixed amount of $NaHCO_3$ (1 mmol). As the initial amount of Mn increased from 2 to 10 mmol, the yield of formic acid increased evidently from 13% to 43% and then remained nearly constant when the amount of Mn reached 8 mmol. The increase in formic acid yield with the increase in Mn most likely occurred because of a stronger reduction condition or a larger amount of hydrogen improving the conversion of CO_2. In addition, it has been reported that formic acid decomposes under high-temperature water (HTW) via dehydration and/or decarboxylation, as shown in Equations 8.3 and 8.4, and the decarboxylation of formic acid is the predominant pathway in HTW (Yasaka et al. 2006; Yu and Savage 1998).

$$HCOOH \leftrightarrow CO + H_2O \tag{8.3}$$

$$HCOOH \leftrightarrow CO_2 + H_2 \tag{8.4}$$

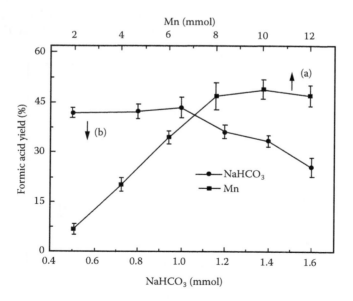

FIGURE 8.3 Effect of the amount of Mn and $NaHCO_3$ on the formic acid yield (temperature: 300°C; time: 2 h; water filling: 35%; $NaHCO_3$: 1 mmol for the effect of the amount Mn; Mn: 8 mmol for the effect of $NaHCO_3$).

Thus, high reductant conditions may also inhibit the decomposition of the formed formic acid. The GC/TCD results confirmed this assumption, which showed that only a small amount of CO_2 and no CO was present in the gas samples.

To examine the effect of the initial amount of $NaHCO_3$ (carbon source) on the formic acid formation from CO_2, experiments were conducted at 300°C for 2 h by fixing the amount of Mn at 8 mmol. As shown in Figure 8.3b, when the amount of $NaHCO_3$ was varied from 0.5 to 1.0 mmol, no obvious change in the yield of formic acid was observed. However, with the amount of $NaHCO_3$ further increased, the formic acid yield decreased very quickly. The formic acid yield having no obvious change in the range of the amount of $NaHCO_3$ between 0.5 and 1 mmol could be explained by the more than 8:1 corresponding ratio of $Mn/NaHCO_3$ as a result of the formic acid yield remaining nearly constant when the ratio of $Mn/NaHCO_3$ was above 8 mmol (see Figure 8.3a). The decrease in the formic acid with a further increase in $NaHCO_3$ could be attributed to a decrease in the ratio of $Mn/NaHCO_3$ to below 8 when $NaHCO_3$ increased to above 1.0 mmol. Therefore, the reactant amounts of 8 mmol Mn and 1 mmol $NaHCO_3$ were chosen in this study.

The effect of the initial solution pH should be an important factor in the reduction of CO_2 to formic acid because pH can affect the decomposition equilibrium of $NaHCO_3$ and the decomposition of formic acid. Some researchers reported that alkaline conditions are generally not favorable for the decomposition of formic acid (Jin et al. 2008). Takahashi et al. (2006) have also reported that formic acid can be selectively formed by CO_2 reduction in a weak alkaline solution under HTW conditions when Fe was used as reductant. Experiments of the effect of pH were conducted by adjusting the initial pH with NaOH or HCl. As shown in Figure 8.4, the highest formic acid yield of 43% occurred at the initial pH of 8.3, which is the same pH value as that observed at 1 mmol $NaHCO_3$ with no additional NaOH or HCl.

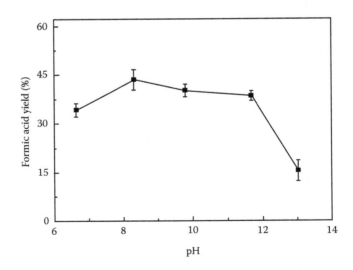

FIGURE 8.4 Effect of pH on the formic acid yield (temperature: 300°C; time: 2 h; $NaHCO_3$: 1 mmol; Mn: 8 mmol; water filling: 35%).

In cases of lower acidity (pH 6.6) and higher alkalinity (pH 13.0), the yield of formic acid decreased to 34% and 10%, respectively. These results demonstrated that a weak alkaline pH value of about 8.3 was favorable for the formation of formic acid, and thus no additional alkali was used in the succeeding experiments.

Subsequently, the effects of reaction temperature and time on the conversion of CO_2 were investigated. As shown in Figure 8.5, when the reaction time increased at 250°C or 275°C, the yield of formic acid was very low and increased inconspicuously. However, the formic acid yield obviously increased when the reaction temperature reached 300°C, and the formic acid yield was as high as 60% when the temperature reached 325°C. Hence, a high temperature was favorable for CO_2 conversion. As shown in Figure 8.4, the time profile indicated that the yield of formic acid increased rapidly at first, and a further increase in reaction time did not result in a significant increase. Therefore, 2 h and 1 h were the optimal reaction times at 300°C and 325°C, respectively.

Finally, water filling is also regarded as an important parameter for CO_2 reduction because water filling can affect the reaction pressure. The effects of water filling were investigated using constant initial amounts of $NaHCO_3$ (1 mmol) and Mn (8 mmol), with the water filling varying from 25% to 55%. As shown in Figure 8.6, the formic acid yield increased evidently with the increase in water filling, and the yields of 76% and 61% were attained when the water filling increased to 55% at 325°C and 300°C, respectively. The results indicated that the increase in water filling was favorable for CO_2 reduction, which may be attributed to the increase in the pressure of the system as water filling increases. Additionally, considering that the increase in water filling led to a decrease in the initial concentration of $NaHCO_3$, which may increase costs for formic acid separation, further experiments were also conducted in different water filling values from 25% to 55% by ensuring that the initial concentration

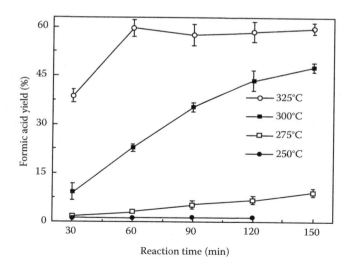

FIGURE 8.5 Effect of temperature and time on the formic acid yield ($NaHCO_3$: 1 mmol; Mn: 8 mmol; water filling: 35%).

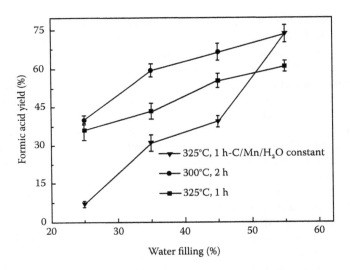

FIGURE 8.6 Effect of water filling on the yield of formic acid (NaHCO$_3$: 1 mmol; Mn: 8 mmol).

of NaHCO$_3$ and Mn is the same as the optimal concentration (1 mmol NaHCO$_3$, 8 mmol Mn with 55% water filling) at 325°C for 1 h. As shown in Figure 8.6, the yield of formic acid clearly increased with the increase in water filling when the ratio of NaHCO$_3$/Mn/H$_2$O was constant, indicating that the increase in formic acid yield was directly related to the pressure but not to the NaHCO$_3$ concentration.

8.3.3 CO$_2$ Role in Improving Hydrogen Production from Water

In our previous study, it was found that no hydrogen was produced in the absence of NaHCO$_3$ for conversion of CO$_2$ under hydrothermal conditions with Fe as reductant; however, a substantial amount of hydrogen was produced in the presence of NaHCO$_3$, which indicated that an increase in the initial NaHCO$_3$ could lead to an increase in hydrogen production. To test whether NaHCO$_3$ also affects hydrogen production when Mn was used, gas samples with and without NaHCO$_3$ were collected and analyzed by GC/TCD. As shown in Table 8.2, in the absence of NaHCO$_3$, 80.5 mL of hydrogen was produced when 4 mmol Mn was used, whereas 89.5 mL of hydrogen was produced in the presence of NaHCO$_3$. Additionally, a formic acid yield of 23% (0.23 mmol formic acid) was obtained in the presence of NaHCO$_3$. Thus, the total hydrogen production in the presence of NaHCO$_3$ was higher than the total hydrogen production without NaHCO$_3$ because the formic acid formed from CO$_2$ hydrogenation consumed some amount of hydrogen. That is to say, CO$_2$ not only acts as a carbon source but also can promote hydrogen generation when Mn was used as reductant. The result is probably caused by the oxidation of Mn shifting the reaction to the direction of formic acid formation owing to the consumption of hydrogen (CO$_2$ hydrogenation) in the presence of CO$_2$, as shown in Equation 8.5. Thus, CO$_2$ provides an additional benefit of improving hydrogen production from water.

TABLE 8.2
Amount of H_2 and CO_2 for Gas Samples and the Yield of Formic Acid

Entry	Mn (mmol)	$NaHCO_3$ (mmol)	H_2 (mL)	CO_2 (mL)	Yield of Formic Acid (%)
1	4	0	80.5	–	–
2	4	1	89.5	0.5	23
3	8	1	140.0	–	43

Note: 300°C; 120 min. Volume of total gas was measured at room temperature (20°C ± 1°C) and At a pressure of 1 atm.

$$Mn + H_2O \longrightarrow \boxed{Mn_xO_y + 2H + HCO_3^- \longrightarrow HCOOH} \quad (8.5)$$

8.3.4 POSSIBLE MECHANISM OF MN OXIDATION IN WATER AND THE ROLE OF MN_xO_y IN THE FORMATION OF FORMIC ACID

Solid residues after reactions were collected and analyzed by XRD to investigate the mechanism of Mn oxidation. Interestingly, the rapid color change of the solid residues from green to brown was observed during collection (see Figure 8.7a). Among all oxides of Mn, the only color of MnO is green, and MnO is easily oxidized to Mn_3O_4. Thus, Mn is most likely oxidized into MnO in the reactions, and the color change is attributed to the further oxidation of MnO during collection in air. As expected, XRD analysis showed that MnO and Mn_3O_4 were detected, as shown in Figure 8.7b, and $Mn(OH)_2$ was also formed. Thus, two pathways for the oxidation of Mn to MnO were possible. One is that MnO is formed directly by the oxidation of Mn, and the other is via the formation of $Mn(OH)_2$, as shown in Equation 8.6.

$$Mn + H_2O \xrightarrow{H_2} MnO \longrightarrow Mn_3O_4$$
$$\searrow \quad \nearrow {-H_2O}$$
$$Mn(OH)_2 + H_2 \quad (8.6)$$

Furthermore, experiments were conducted using short reaction times of 1, 5, and 60 min to determine the reaction pathway of the oxidation of Mn into MnO. As shown in Figure 8.8, the solid with no green color was observed during the collection of solid samples, and XRD analysis showed that only $Mn(OH)_2$ was formed when the reaction times were 1 and 5 min. As the reaction time increased to 60 min, MnO and Mn_3O_4 were observed, whereas the amount of $Mn(OH)_2$ and Mn decreased

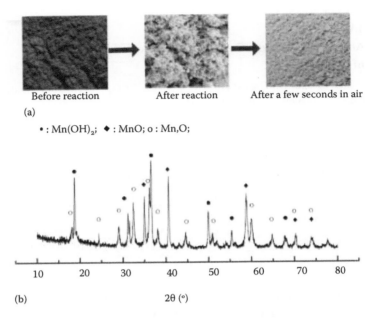

FIGURE 8.7 (a) Photographs of solid samples; (b) XRD patterns of the solid samples (NaHCO$_3$: 1 mmol; time: 2 h; Mn: 8 mmol; temperature: 300°C; water filling: 35%).

gradually. Simultaneously, the color change from green to brown in a solid sample was observed during the collection of solid samples. These results suggested that the oxidation of Mn to MnO occurs via the formation of Mn(OH)$_2$.

It is generally known that a catalyst is needed for activating the hydrogen in the hydrogenation of CO$_2$. However, interestingly, there was a high formic acid yield without any catalyst added in the present study, which suggested that some intermediates such as Mn$_x$O$_y$ formed in situ may act as a catalyst in the reaction. To investigate this topic, experiments with NaHCO$_3$ and gaseous hydrogen were conducted by changing the amount of gaseous hydrogen from 5 to 18 mmol. As shown in Figure 8.9, all of the formic acid yields with gaseous hydrogen were very low, keeping in only about 2%, while a considerably high yield of formic acid can be obtained when Mn was used to react with water to produce hydrogen. These results suggested that Mn$_x$O$_y$ may act as a catalyst in the reduction of CO$_2$ to formic acid with Mn. To further provide evidence, experiments with gaseous hydrogen and MnO or Mn$_3$O$_4$ additive as catalyst were conducted. As shown in Figure 8.9, the formic acid increased to 9% when MnO was added (Run 2); however, the formic acid was less than 2% (Run 3) when Mn$_3$O$_4$ was added. The results indicated that MnO can provide a catalytic activity in the reduction of CO$_2$ into formic acid and Mn$_3$O$_4$ cannot. However, the yield of 9% with MnO was much lower than that with Mn. One of the possible explanations is that the catalytic activity of MnO formed in situ is much more than the added MnO.

Hydrothermal CO_2 Reduction with Manganese to Produce Formic Acid

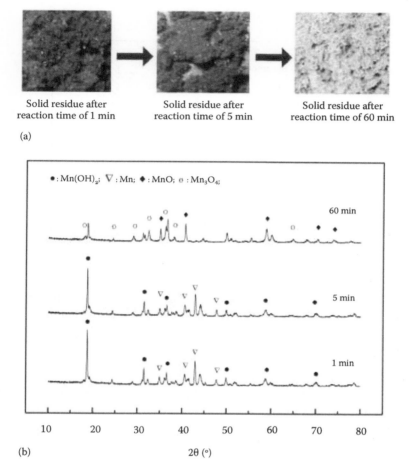

FIGURE 8.8 (a) Photographs of solid samples; (b) XRD patterns of the solid sample ($NaHCO_3$: 1 mmol; Mn: 8 mmol; temperature: 300°C; water filling: 35%).

8.3.5 Investigation of Formic Acid Formation via HCO_3^- or Gaseous CO_2

After understanding the pathway of Mn oxidation promotion of hydrogen production in the presence of $NaHCO_3$, we also wanted to know the mechanism for formic acid formation, that is, whether its formation pathway is via gaseous CO_2 or HCO_3^-. To achieve this objective, experiments with $NaHCO_3$ and gaseous CO_2 in the presence of Mn were conducted. As shown in Table 8.3, a formic acid yield of 43% was achieved when 1 mmol $NaHCO_3$ was used (entry 1). However, when using gaseous CO_2 as a reactant instead of $NaHCO_3$ in the absence of NaOH and in the presence of NaOH

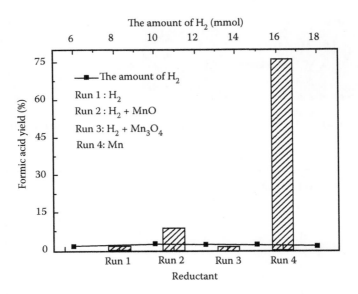

FIGURE 8.9 Formic acid yields obtained with different amount of gaseous H_2 and additive MnO or Mn_3O_4 (325°C; 1 h; $NaHCO_3$: 1 mmol; water filling: 55%; Mn: 8 mmol; gaseous H_2: 6 mmol; MnO: 6 mmol; Mn_3O_4: 6 mmol).

TABLE 8.3
Difference in the Yields of Formic Acid between $NaHCO_3$ and Gaseous CO_2

Entry	$NaHCO_3$ (mmol)	CO_2 (mmol)	NaOH (mmol)	H_2O (mL)	pH	Yield (%)
1	1	–	–	2	8.6	43
2	–	1	–	2	6.7	0.1
3	–	1	1	2	13.6	5.5
4	1	–	–	–	–	0.2

Note: Mn: 8 mmol; temperature: 300°C; time: 120 min.

(entries 2 and 3), 0.2% and 5.5% formic acid yields were obtained, respectively, suggesting that the formation of formic acid is closely related to pH; namely, it is related to the concentration of CO_2 in the solvent, and the low pH is not favorable for the dissolution of CO_2, which is not favorable for the formic acid formation. If this hypothesis is true, the dissolution time of CO_2 before the reactions increase should lead to an increase in formic acid yield. To validate this hypothesis, experiments were performed with an additional dissolution process at room temperature before the reactor was put into a salt bath in an attempt to increase the dissolution of CO_2 in NaOH solution. As expected, the initial pH of the solution clearly decreased with

Hydrothermal CO₂ Reduction with Manganese to Produce Formic Acid

the prolongation of dissolution time and the yield of formic acid increased evidently, as shown in Figure 8.10, and the best formic acid yield was achieved at a weak alkaline pH value. This observation is in agreement with the fact discussed in Figure 8.4 that the best formic acid yield was achieved in a weak alkaline pH. CO_2 can exist as different forms of hydrogen carbonate and carbonate at different pH, as shown in Equation 8.7, and the distribution coefficient of HCO_3^- is more than 0.9 when the pH value is 8.3, as shown in Equation 8.8. As discussed before, the highest formic acid yield occurred at an initial pH of 8.3. The yield of formic acid significantly decreased under more acidic or more alkaline conditions, which may be attributed to the fact that the main ion present is not HCO_3^- when the pH is less than 6.3 or more than 10.3. These results are further evidence that formic acid formation from water splitting reduction of CO_2 with Mn occurs via HCO_3^- rather than via CO_2.

$$CO_2 + H_2O \underset{pK_1=6.3}{\rightleftharpoons} HCO_3^- + H^+ \underset{pK_2=10.3}{\rightleftharpoons} CO_3^{2-} + 2H^+ \quad (8.7)$$

$$pH = p\sqrt{\frac{K_2[HCO_3^-] + K_w}{1 + \frac{[HCO_3^-]}{K_1}}} = p\sqrt{K_2 * K_1} = \frac{1}{2}(pK_1 + pK_2) = 8.3 \quad (8.8)$$

Furthermore, applying the same thermodynamics method, we calculated the reaction heat and free energy for the formic acid production at 298K.

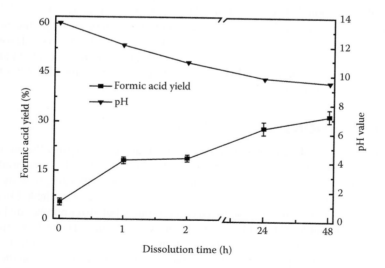

FIGURE 8.10 Yield of formic acid with the dissolution of CO_2 (Mn: 8 mmol; CO_2: 1 mmol; NaOH: 1 mmol; water filling: 35%; dissolution of CO_2: at room temperature; reaction conditions: temperature: 300°C; time: 2 h).

$$\Delta G = G_{\text{product}} - G_{\text{reactants}} \tag{8.9}$$

Thus, the energy required for the producing formic acid via H_2 and H was as follows:

$$HCO_3^- + H_2 \rightleftharpoons HCOO^- + H_2O \quad \Delta G_{298}^o = -7 \text{ kJ/mol}$$
$$\Delta H_{298}^o = +24.62 \text{ kJ/mol} \tag{8.10}$$

$$CO_2 + H_2 \rightleftharpoons HCOOH + \quad \Delta G_{298}^o = +33 \text{ kJ/mol}$$
$$\Delta H_{298}^o = -31.21 \text{ kJ/mol} \tag{8.11}$$

$$HCO_3^- + 2H \rightleftharpoons HCOO^- + H_2O \quad \Delta G_{298}^o = +399.48 \text{ kJ/mol}$$
$$\Delta H_{298}^o = -411.35 \text{ kJ/mol} \tag{8.12}$$

$$CO_2 + 2H \rightleftharpoons HCOOH \quad \Delta G_{298}^o = -373.48 \text{ kJ/mol}$$
$$\Delta H_{298}^o = -467.18 \text{ kJ/mol} \tag{8.13}$$

From a thermodynamic point of view, the standard free energy change for CO_2 hydrogenation with H into formic acid in the aqueous phase is −399.48 kJ/mol, whereas the reaction between H and CO_2 in the gas phase is −373.48 kJ/mol (Equations 8.12 and 8.13). Compared with calculation, it can be seen that the energy of formic acid formation via H is much negative than via H_2, which is more favorable for the conversion of CO_2 to formic acid. These calculation results indicated that the reaction process is exothermic and HCO_3^- is preferred over CO_2 for the interconversion between hydrogen and formic acid.

8.4 CONCLUSIONS AND PROSPECTS

In summary, we have developed the first no-catalyst-added and highly efficient water splitting for the highly selective conversion of CO_2 into formic acid, which produces a considerably high formic acid yield of more than 75% with an exceptionally high selectivity of more than 98% from CO_2 using simple, commercially available Mn powder. In this reaction, CO_2 not only acts as a carbon source but also improves hydrogen production from water. The oxidation of Mn to MnO occurs via the formation of $Mn(OH)_2$, and MnO can provide catalytic activity in the reduction of CO_2 into formic acid. The proposed processes can be linked to provide a promising method for improving artificial photosynthetic efficiency by water splitting for CO_2 reduction combined with the MnO/Mn cycle using solar energy.

REFERENCES

Cornia A., Caneschi A., Dapporto P. et al. 1999. Manganese(III) formate: A three-dimensional framework that traps carbon dioxide molecules. *Angewandte Chemie-International Edition, 38*: 1780–1782.

Cox P. M., Betts R. A., Jones C. D., Spall S. A., Totterdell, I. J. 2000. Acceleration of global warming due to carbon-cycle feedbacks in a coupled climate model. *Nature 408*: 184–187.

Galvez M. E., Loutzenhiser P. G., Hischier I., Steinfeld A. 2008. CO_2 splitting via two-step solar thermochemical cycles with Zn/ZnO and FeO/Fe_3O_4 redox reactions: Thermodynamic analysis. *Energy & Fuels 22*: 3544–3550.

Grimm N. B., Chapin F. S., Bierwagen, B. et al. 2013. The impacts of climate change on ecosystem structure and function. *Frontiers in Ecology and the Environment 11*: 474–482.

Haueter P., Moeller S., Palumbo R., Steinfeld A. 1999. The production of zinc by thermal dissociation of zinc oxide—Solar chemical reactor design. *Solar Energy 67*: 161–167.

Hu Y. H. 2012. A highly efficient photocatalyst hydrogenated black TiO_2 for the photocatalytic splitting of water. *Angewandte Chemie-International Edition 51*: 12410–12412.

Indrakanti V. P., Kubicki J. D., Schobert H. H. 2009. Photoinduced activation of CO_2 on Ti-based heterogeneous catalysts: Current state, chemical physics-based insights and outlook. *Energy & Environmental Science 2*: 745–758.

Jessop P. G., Joo F., Tai C. C. 2004. Recent advances in the homogeneous hydrogenation of carbon dioxide. *Coordination Chemistry Reviews 248*: 2425–2442.

Jin F., Gao Y., Jin Y. et al. 2011. High-yield reduction of carbon dioxide into formic acid by zero-valent metal/metal oxide redox cycles. *Energy & Environmental Science 4*: 881–884.

Jin F., Kishita A., Moriya T. et al. 2001. Kinetics of oxidation of food wastes with H_2O_2 in supercritical water. *Journal of Supercritical Fluids 19*: 251–262.

Jin F., Yun J., Li G. et al. 2008. Hydrothermal conversion of carbohydrate biomass into formic acid at mild temperatures. *Green Chemistry 10*: 612–615.

Jin F., Zhou Z. Y., Moriya T. et al. 2005. Controlling hydrothermal reaction pathways to improve acetic acid production from carbohydrate biomass. *Environmental Science & Technology 39*: 1893–1902.

Kone A. C. and Buke T. 2010. Forecasting of CO_2 emissions from fuel combustion using trend analysis. *Renewable & Sustainable Energy Reviews 14*: 2906–2915.

Meng M. and Niu D. 2011. Modeling CO_2 emissions from fossil fuel combustion using the logistic equation. *Energy 36*: 3355–3359.

Neelameggham N. R. 2008. Solar pyrometallurgy—An historic review. *JOM 60*: 48–50.

Neelameggham N. R. 2009. Soda-fuel metallurgy: Metal ions for carbon neutral CO_2 and H_2O reduction. *JOM 61*: 25–27.

Park A.-H. A., Fan L.-S. 2004. CO_2 mineral sequestration: Physically activated dissolution of serpentine and pH swing process. *Chemical Engineering Science 59*: 5241–5247.

Patz J. A., Campbell-Lendrum D., Holloway T., Foley J. A. 2005. Impact of regional climate change on human health. *Nature 438*: 310–317.

Roy S. C., Varghese O. K., Paulose M., Grimes C. A. 2010. Toward solar fuels: Photocatalytic conversion of carbon dioxide to hydrocarbons. *ACS Nano 4*: 1259–1278.

Shabala S. 2009. Metal cations in CO_2 assimilation and conversion by plants. *JOM 61*: 28–34.

Sheshko T. F. and Serov, Y. M. 2012. Bimetallic systems containing Fe, Co, Ni, and Mn nanoparticles as catalysts for the hydrogenation of carbon oxides. *Russian Journal of Physical Chemistry A 86*: 283–288.

Sumaila U. R., Cheung W. W. L., Lam V. W. Y., Pauly. D., Herrick S. 2011. Climate change impacts on the biophysics and economics of world fisheries. *Nature Climate Change 1*: 449–456.

Takahashi H., Liu L. H., Yashiro Y. et al. 2006. CO_2 reduction using hydrothermal method for the selective formation of organic compounds. *Journal of Materials Science 41*: 1585–1589.

Takeda H., Koike K., Inoue H., Ishitani O. 2008. Development of an efficient photocatalytic system for CO_2 reduction using rhenium(l) complexes based on mechanistic studies. *Journal of the American Chemical Society 130*: 2023–2031.

Tian G., Yuan H., Mu Y., He C., Feng S. 2007. Hydrothermal reactions from sodium hydrogen carbonate to phenol. *Organic Letters 9*: 2019–2021.

Uchida S. 2012. Solar power concentration using laser technology for the magnesium energy circulation. *Solar Paces* 1–11.

Varghese O. K., Paulose M., LaTempa T. J., Grimes C. A. 2009. High-rate solar photocatalytic conversion of CO_2 and water vapor to hydrocarbon fuels. *Nano Letters 9*: 731–737.

Wang W., Wang, S., Ma X., Gong J. 2011. Recent advances in catalytic hydrogenation of carbon dioxide. *Chem Soc Rev 40* (7): 3703–3727.

Wheeler T., Braun J. 2013. Climate change impacts on global food security. *Science 341*: 508–513.

Wu B., Gao Y., Jin F. et al. 2009. Catalytic conversion of $NaHCO_3$ into formic acid in mild hydrothermal conditions for CO_2 utilization. *Catalysis Today 148*: 405–410.

Yadav R. K., Baeg J.-O., Park N. J. et al. 2012. A photocatalyst-enzyme coupled artificial photosynthesis system for solar energy in production of formic acid from CO_2. *Journal of the American Chemical Society 134*: 11455–11461.

Yasaka Y., Yoshida K., Wakai C., Matubayasi N., Nakahara M. 2006. Kinetic and equilibrium study on formic acid decomposition in relation to the water-gas-shift reaction. *Journal of Physical Chemistry A 110*: 11082–11090.

Yu J. L. and Savage P. E. 1998. Decomposition of formic acid under hydrothermal conditions. *Industrial & Engineering Chemistry Research 37*: 2–10.

9 Autocatalytic Hydrothermal CO_2 Reduction with Aluminum to Produce Formic Acid

Binbin Jin, Guodong Yao, Fangming Jin, and Heng Zhong

CONTENTS

9.1 Introduction 127
9.2 Experimental Section 130
    9.2.1 Materials 130
    9.2.2 Experimental Procedure 130
    9.2.3 Product Analysis 131
    9.2.4 Quantum Chemical Calculations 131
9.3 Results and Discussion 131
    9.3.1 Hydrogen Production by Water Splitting with Al 131
    9.3.2 Hydrogenation of CO_2 by Water Splitting with Al 132
    9.3.3 Waste Metal as a Reducing Agent for Hydrogenation of CO_2 135
    9.3.4 Proposed Mechanism of the Reduction of CO_2 by Water Splitting with Al 136
9.4 Conclusions 138
Acknowledgments 138
References 138

9.1 INTRODUCTION

Nowadays, developing clean and renewable energy to replace traditional fossil fuels has become a major research topic. Human activities produce an excess amount of CO_2 to the carbon cycle owing to the consumption of fossil energy, and the increased content of CO_2 has caused a series of environmental problems affecting sustainable development. Therefore, for sustainable development, utilization of resources, environmental protection, and developing efficient approaches to decrease CO_2 are necessary, and various methods have been proposed. Notably, utilizing CO_2 as a carbon source for organic chemicals production would be an especially promising technology [1,2].

Among these methods, photocatalytic conversion of CO_2 is quite attractive to us. As a clean energy, it is very ideal to use solar energy to drive the conversion process of CO_2 into chemicals and fuel energy. Therefore, extensive attempts have been made toward solar-to-fuel conversion by artificial photosynthetic conversion of CO_2 [3–6]. However, high efficient conversion with direct use of solar energy is still faced with great challenges. Compared to direct solar-to-fuel conversion, the integrated technology that can utilize solar energy indirectly is expected to have the potential to improve the efficiency of artificial photosynthesis. The electrochemical reduction of CO_2 is considered as a typical integrated technology to increase artificial photosynthetic efficiency because electricity can be generated from solar energy [7]. Furthermore, other efficient integrated technologies have been studied recently. For example, Steinfeld reported CO_2 splitting via two-step solar thermochemical cycles based on metal/metal oxide redox reactions. However, the product was limited to CO, and hydrogen from H_2O cannot be used in the reaction [8–10].

Currently, the catalytic conversion of CO_2 with the addition of hydrogen is regarded as the most commercially feasible method, and this research area has become increasingly active in both fundamental and industrial applications. However, gaseous hydrogen is an energy-intensive material that is mainly produced by the reformation of hydrocarbons. Meanwhile, hydrogen storage also needs energy. Therefore, the supply of economical and safe gas hydrogen sources is considered as the major bottleneck and poses significant challenges. Thus, a highly efficient and simple process for CO_2 reduction to chemicals and fuels is highly desired. The conversion of dissolved CO_2 into fossil fuels in the Earth's crust inspires us to mimic hydrocarbon formation in nature. The abiotic synthesis of organics suggests that highly efficient dissociation of H_2O and subsequent reduction of CO_2 into organics could be achieved with metals under hydrothermal conditions. In our previous contributions, we have investigated the reduction of CO_2 with various zero-valence metals, such as Fe and Zn. CO_2 reduction with Fe, Zn, Al, and Mn under hydrothermal conditions was investigated and formic acid formed in all cases. In these reaction systems, water can be used as an in situ liquid hydrogen source and then hydrogen storage, transportation, and purity are not required. The relative efficiencies of the metals for converting CO_2 into formic acid, defined as the molar ratio of formic acid/metal, were Al > Zn > Mn > Fe, which was almost the same as their activities. Fe has shown potential for the dissociation of H_2O for CO_2 hydrogenation. However, the yield of the product, formic acid, was relatively low, even with the addition of a nickel catalyst, and the highest formic acid yield on a carbon basis was approximately 16% [11]. Though Zn exhibits more excellent reduction ability for hydrogenation of CO_2 (~80% yield of formic acid) [12], the content of Zn in Earth is low (~0.007% in mass). Al is the most abundant metal in the crust (8% in mass) and possesses excellent reducing ability for displacing hydrogen from H_2O. If CO_2 could be reduced efficiently into formic acid with Al under hydrothermal conditions, a novel facile method for hydrogenation of CO_2 based on the abundant metal reduction will be developed. Furthermore, the Al oxidation product can also be reduced to Al similar to the reduction of ZnO using concentrated solar energy, which has been well studied [8–10] and applied on an industrial scale [13,14]. Hence, by combining technologies that use solar energy to reduce Al oxide to Al, Al materials circulation can be achieved in the reduction of CO_2 with Al under hydrothermal conditions, as shown in Scheme 9.1.

In the whole process, the reduction of CO_2 with Al under hydrothermal conditions is exothermic, and the reduction of Al oxide to Al can occur with solar energy. Thus, in this context, rapidly and efficiently solar-derived water splitting for the reduction of CO_2 to chemicals will be achieved. With the consideration of this concept, we investigated the possibility of hydrogen production by water splitting with Al and in situ reduction of CO_2, and high-temperature water (HTW) was also applied because the hydrothermal condition could improve the mass and heat transition, accelerating the hydrogenation of CO_2. Hydrothermal chemistry has received much attention in organic chemical synthesis and biomass conversion because of the unique inherent properties of HTW, including a high ion product, a low dielectric constant, a fast reaction rate, and an environmentally benign property. We have conducted much research involving biomass conversion under hydrothermal reactions [15]. On the other hand, hydrothermal reactions have played an important role in the formation of fossil fuel in the Earth's crust and deep-sea hydrothermal vent such as the abiotic synthesis of hydrocarbons from the dissolved CO_2 [16]. Thus, CO_2 reduction in the simulated hydrothermal vent system should be a very desirable research. Additionally, water also played a role as hydrogen source.

Furthermore, it has been reported that individual (and neutral) Al atoms could react with a few water molecules to produce HAl-OH·nH_2O or AlOH·nH_2O + H both in the gas phase and within clusters of water molecules with almost no energy barrier [17]. However, until now, detailed information on the reaction kinetics of Al with H_2O is limited; even McClean reported mechanisms to describe the combustion of aluminum particles utilizing the rate constant of the reaction between Al atom and H_2O molecule [18]. In particular, for the highly efficient reduction of CO_2 to formic acid with metal under hydrothermal conditions, a detailed mechanism is still unknown because the reaction intermediates are difficult to experimentally identify under hydrothermal conditions, for example, via an in situ technique, owing to the reactions taking place in HTW.

Density functional theory (DFT) has always been used in the theoretical study of various types of chemical reactions, including the reaction of Al with H_2O [19,20].

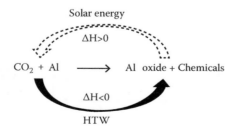

SCHEME 9.1 Proposed carbon cycle with Al as a reductant under hydrothermal conditions combined with a reduction of Al oxide into Al system using solar energy.

Additionally, in our previous study, we also reported that zinc hydride (Zn-H) is a key intermediate species in the reduction of CO_2 to formic acid with Zn under hydrothermal conditions, which demonstrates that the formation of formic acid is through an SN2-like mechanism [21]. Thus, in this study, for a better understanding of the reaction pathway and scheme, a theoretical study was performed to simulate the reaction pathway based on our experimental results.

9.2 EXPERIMENTAL SECTION

9.2.1 MATERIALS

$NaHCO_3$ (AR, 98%), Al (powder, 200 mesh), and formic acid (AR, 98%) were purchased from Sinopharm Chemical Reagent Co., Ltd. Deionized water was used in the study.

9.2.2 EXPERIMENTAL PROCEDURE

With the consideration that CO_2 dissolved in water produces HCO_3^- in nature, $NaHCO_3$ was used as a CO_2 source in our study. Furthermore, salts containing HCO_3^- are products formed by carbon capture and storage technology. The schematic drawing of the experimental equipment is shown in Figure 9.1 and a detailed description can be found elsewhere [22,23]. In a typical procedure, the desired amount of $NaHCO_3$

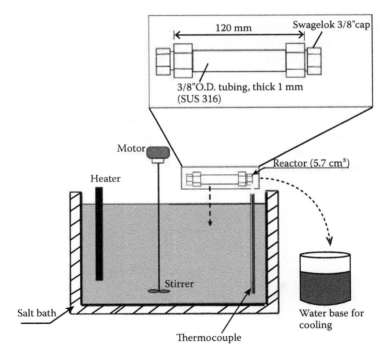

FIGURE 9.1 Schematic of batch reactor system.

(CO_2 source), reductant (Al powder), and 2 mL of deionized water were loaded in a batch reactor (~35% water filling). The batch reactor was 3/8 in. stainless steel SUS 316 tubing with fittings (Swagelok, SUS 316) sealed at each end. The reactor had a total length of 120 mm, a wall thickness of 1 mm, and an inner volume of 5.7 mL. The reactor was connected to a gas valve for gas collection. After loading, the reactor was immersed in a salt bath, which had been filled with $NaNO_3$ and KNO_3 salts mixed at a ratio of ca. 1:1 and preheated to the desired temperature (250–325°C). During the reaction, the reactor was shaken while being kept horizontally. After the reaction, the reactor was removed from the salt bath to quench in a cold water bath. After cooling the reactor down to room temperature, the valve was opened and the gas sample was collected by a graduated cylinder in the tank that was filled with saturated NaCl solution.

9.2.3 Product Analysis

The liquid products were measured by two detectors (UV detector and refractive index detector) equipped with Agilent 1200 high-performance liquid chromatography (HPLC) with one Shodex RSpak KC-G and two RSpak KC-811 columns, using 2 mmol/L $HClO_4$ solution as flowing solvent. The total inorganic carbon was analyzed by a Shimadzu TOC 5000A. Gas samples were analyzed by GC-TCD, and solid residues were measured by x-ray diffraction. Quantitative analyses of formic acid were based on the average value obtained from two sample analyses with the relative errors always less than 5% for all experiments. Solid samples were analyzed by x-ray diffraction (Bruker D8 Advance).

Formic acid yield was defined as the percentage of formic acid to the initial $NaHCO_3$ based on the carbon basis. The selectivity of formic acid was defined as the percentage of carbon contained in formic acid to the total organic carbon in the liquid phase.

9.2.4 Quantum Chemical Calculations

The strategy consists of a series of ab initio and DFT quantum chemical calculations for the reaction. All the structures at equilibrium and transition states were optimized by a B3LYP functional using the extended 6-311 + G(3df,2p) basis set. To take the solvation effect into account, the polarizable continuum model was applied. The transition state was calculated by following minimum energy paths of the reactions with the use of the Gonzalez–Schlegel intrinsic reaction coordinate algorithm. All calculations were performed using Gaussian 09 [24].

9.3 RESULTS AND DISCUSSION

9.3.1 Hydrogen Production by Water Splitting with Al

First, it is notable that hydrogen production is key to the hydrogenation reduction of CO_2 into formic acid at the current conditions. Thus, experiments were first conducted to examine whether Al could be employed as a reducing agent for water

TABLE 9.1
Effect of NaHCO₃ Amount on the Gas Production

Metal	NaHCO₃ (mmol)	H₂ (mL)	CO₂ (mL)	Total Gas (mL)
Al	0	114	0	114
	1	122.5	5.5	128
	2	121.5	16.5	138

splitting under hydrothermal conditions. First, the effect of initial pH on hydrogen production was investigated. In our previous study, we discovered that Fe cannot displace hydrogen from H_2O in neutral solution at 300°C for 2 h. To investigate the hydrogen production by water splitting with Al, experiments were conducted at pH 7 and 8.6 at 300°C. The results showed that 4 mmol Al can displace 58 mL of hydrogen in the neutral solution after 2 h. When the pH value was adjusted to weak alkalinity with NaOH, the amount of hydrogen increased sharply to 114 mL. Obviously, a higher alkalinity is favorable for hydrogen production. It is probably that OH^- can promote the Al oxidation in solution.

The effect of $NaHCO_3$ concentration on hydrogen production was then studied. The pH value of the $NaHCO_3$ solution is approximately 8.6; thus, $NaHCO_3$ should also promote hydrogen production with Al in HTW. Experimental results fit this deduction and the amount of hydrogen increase with 1 mmol $NaHCO_3$. However, when the amount of $NaHCO_3$ increased to 2 mmol, the amount of hydrogen changes slightly (Table 9.1). Therefore, based on the experimental results, 1 mmol $NaHCO_3$ is suitable for hydrogen production at 300°C.

9.3.2 Hydrogenation of CO_2 by Water Splitting with Al

As mentioned above, $NaHCO_3$ can promote hydrogen production by water splitting with Al. To examine the possibility of hydrogenation of CO_2 with the hydrogen formed in situ, liquid samples were analyzed by HPLC. As shown in Figure 9.2, only formic acid peak was observed, indicating no other liquid product formation. TOC analysis showed the organic carbon mainly exists in formic acid; that is, the selectivity of formic acid on the carbon basis is close to 100%. Figure 9.3 shows XRD patterns of the solid samples after reactions. Whether the solid samples were collected after 30 or 90 min, all peaks belong to AlO(OH), indicating that Al was completely oxidized during the reaction.

To determine the optimum conditions for the hydrogenation of $NaHCO_3$, effects of the Al loading amount, $NaHCO_3$ loading amount, reaction time, and reaction temperature on the yield of formic acid were investigated. First, the effect of Al amount on the yield of formic acid was studied by varying the Al amount from 1 to 8 mmol with 1 mmol $NaHCO_3$ and 2 mL of H_2O at 300°C for 2 h. As shown in Figure 9.4, the yield of formic acid increased significantly with the increase of Al loading amount. When the Al amount increased to 6 mmol, 62.8% yield of formic acid was obtained. However, excess Al (more than 8 mmol) is not favorable for formic acid formation. It is likely that Al oxide would also promote the decomposition of formic acid,

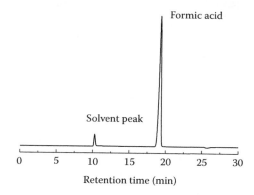

FIGURE 9.2 HPLC chromatogram of liquid sample after the reaction of 1 mmol NaHCO$_3$ and 6 mmol Al in 2 mL H$_2$O at 300°C for 2 h.

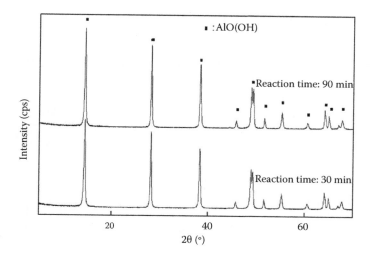

FIGURE 9.3 XRD patterns of the solid samples after reactions at 300°C.

decreasing the formic acid yield. Thus, based on the results, 6 mmol Al is suitable for the conversion of NaHCO$_3$ in this study.

Then, the effect of NaHCO$_3$ amount on the yield of formic acid was studied by varying the NaHCO$_3$ amount from 1 to 8 mmol with 6 mmol Al at 300°C for 2 h. As illustrated in Figure 9.4, the formic acid yield decreased gradually with the NaHCO$_3$ increase. It could be that no sufficient amount of Al can reduce excess NaHCO$_3$. Thus, the amount of NaHCO$_3$ was set at 1 mmol during the reaction.

Next, the influence of reaction time on the yield of formic acid was studied by changing the reaction time from 1 to 2 h with 6 mmol Al and 1 mmol NaHCO$_3$ at 300°C. As shown in Figure 9.5, the yield of formic acid increases with the reaction time increase. A higher yield of formic acid, 64%, was obtained when the reaction

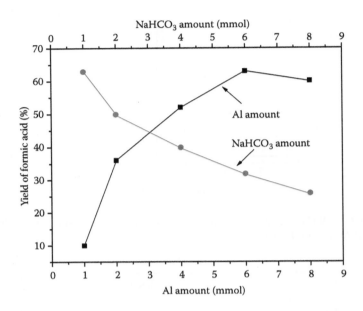

FIGURE 9.4 Effects of Al and $NaHCO_3$ amounts on the yield of formic acid (300°C, 2 h).

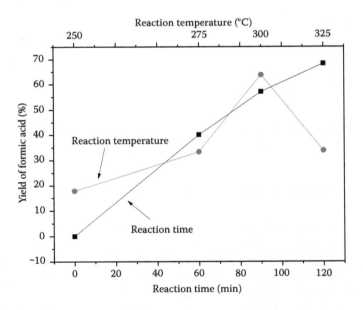

FIGURE 9.5 Effects of reaction time and temperature on the yield of formic acid (6 mmol Al, 1 mmol $NaHCO_3$).

time was 2 h. Although a higher yield of formic acid will be achieved with prolonging the reaction time, 2 h is better considering the cost and reactor safety.

Finally, experiments were performed to examine the effect of reaction temperature on the yield of formic acid at different temperatures from 250°C to 300°C for 2 h. The results showed that the yields of formic acid remarkably increased as the temperature increased from 250°C to 300°C (Figure 9.5). A 64% yield was obtained at 300°C. However, interestingly, the yield of formic acid decreased to 34% at 325°C. The reason for this is probably that the decomposition of formic acid speeds up at a higher temperature. Therefore, the preferred reaction temperature is 300°C.

9.3.3 Waste Metal as a Reducing Agent for Hydrogenation of CO_2

The bulk Al powder shows a high activity for CO_2 hydrogenation under hydrothermal conditions. However, direct use of pure Al is not economical. Discarded electrical equipment and cars, for example, contain abundant Al and other metals. If we can use these waste metals as a reducing agent for the hydrogenation of CO_2, there will be a significant cost reduction. Furthermore, the utilization of the waste metals is also achieved. To assess the possibility of waste metals as a reducing agent, metals with various ratios according to typical waste composition and 7.5 mmol $NaHCO_3$ were added into a larger SUS 316 stainless steel batch reactor with an inner volume of 42 mL. As illustrated in Table 9.2, the satisfactory yields of formic acid were obtained. Additionally, H_2 collected in the gas was more than 90%. Thus, CO_2 reduction by waste metals is feasible.

TABLE 9.2
Yields of Formic Acid with Different Metals (Experimental Conditions: Reaction Temperature: 300°C; Reaction Time: 2 h; $NaHCO_3$: 7.5 mmol; Water Filling: 35%)

Zn (mmol)	Al (mmol)	Fe (mmol)	Cu (mmol)	Sn (mmol)	Ni (mmol)	Y_{HCOOH} (%)	Gas (mL)
1.38	33.3	26.8	7.1	0.505	–	60	1080
6	–	2	–	–	–	56.9	–
6	–	–	–	–	6	31.5	130
10	–	–	–	–	–	75.1	120
–	5	1	–	–	–	44.5	–
–	6	–	–	–	3	31.2	215
–	6	–	–	–	–	62.8	190

9.3.4 Proposed Mechanism of the Reduction of CO_2 by Water Splitting with Al

To date, a detailed reaction mechanism of Al conversion in HTW is still ambiguous. Sharipov et al. have studied the Al reaction with H_2O at 298–1174 K by using experimental and theoretical studies [25]. The theoretical study presented the reaction pathway leading to the aluminum hydrate formation in trans isomeric to be twofold degenerate. Then, the HAlOH^{trans} structure can transform into the HAlOH^{cis} structure with its internal rotation, that is, Al + H_2O → HAlOH^{trans} → HAlOH^{cis}. Hauge et al.'s experimental results confirmed that a HAlOH adduct or a more complex system involving more water molecules is formed in the co-condensation of Al atoms with water [26]. It was also interpreted by Joly et al. with condensation and EPR spectroscopic experiments [27]. Douglas et al. also suggested that HAlOH is formed for its recorded electronic spectrum [28]. More interestingly, in the study by Barcia and Flores, they thought that water molecules may be involved with Al molecules, which would mean that they act as a catalyst of the Al + H_2O → HAlOH process, and HAlOH could be the key intermediate in the formation of hydroxides Al(OH)$_m$ (m = 1–3) by individual Al atoms, through the elimination of atomic or molecular hydrogen [29]. Al(OH)$_3$ could be converted into AlO(OH) through the dehydration process at 300°C.

Although we cannot observe HAlOH by XRD and ATR-FTIR, we believe that HAlOH is a key intermediate in our experiments. It should be noted that the Al–H bond in HAlOH is strongly polarized with Al positive and H negative. This abnormal polarization of Al–H leads to high reactivity, and it is also crucial for HCOO$^-$ formation. Therefore, at first, the formation of HAlOH species, an intermediate in the reaction of Al with H_2O, was calculated. The structures of Al + H_2O, HAlOH, and transition states were optimized by the B3LYP functional using the extended 6-311 + G(3df,2p) basis set. For the HAlOH structure, we have carried out an optimization of bond length and angles. The geometries and charge could be seen in Figure 9.6. As shown in Figure 9.6, the calculated Al–O distance of initial state is 2.079 Å, whereas the transition state and final state are 1.850 and 1.718 Å. The results demonstrated that with the reaction taking place, OH$^-$ is becoming closer with Al, along with the formation of Al–H species. In the initial state, the Mulliken charge of H is positive. However, in the transition state and final state, the charge of one H atom became negative. In the final state, the H atom with negative charge is very close with the Al atom, forming metal hydrides. The calculated Al–H distance of the final state is 1.609 Å, which is larger than the distance of O–H. The calculated energy results showed that the HAlOH species is more stable (with respect to Al + H_2O) by −31.96 kcal/mol. The calculated energy of the transition state is larger (with respect to Al + H_2O) by 18.18 kcal/mol, which is still a small value. Considering that this reaction took place under mild hydrothermal conditions, the reaction can easily occur. It should be noted that HAlOH^{cis} was discussed only in this study, because the condition of HAlOH^{trans} is similar.

Based on the above results and discussion, a possible mechanism of hydrogenation of NaHCO$_3$ by water splitting with Al was proposed, as shown in Scheme 9.2. First, aluminum hydrate, HAlOH, was formed in the reaction of Al with H_2O.

Hydrothermal CO$_2$ Reduction with Aluminum to Produce Formic Acid

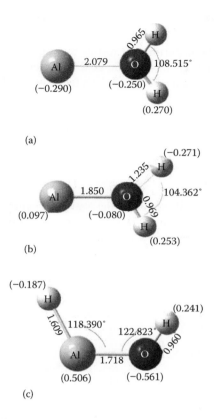

FIGURE 9.6 The geometries and charge of initial state (a), TS (b), and final state (c) of the reaction Al + H$_2$O → HAl(OH).

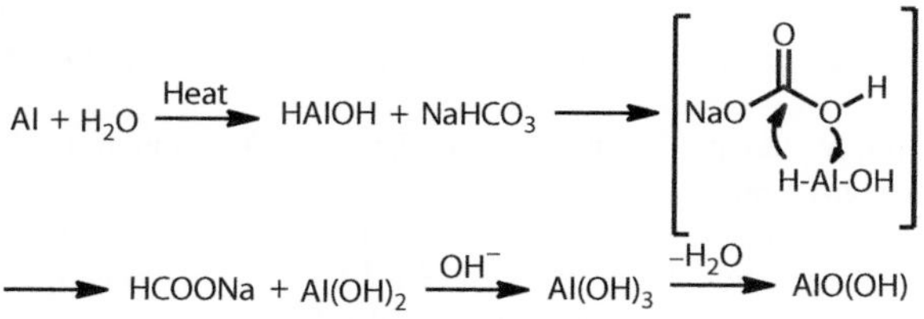

$$Al + H_2O \xrightarrow{Heat} HAlOH + NaHCO_3 \longrightarrow \left[NaO\underset{H-Al-OH}{\overset{O}{\diagdown}}O\diagdown H \right]$$

$$\longrightarrow HCOONa + Al(OH)_2 \xrightarrow{OH^-} Al(OH)_3 \xrightarrow{-H_2O} AlO(OH)$$

SCHEME 9.2 Proposed mechanism of hydrogenation of NaHCO$_3$ by water splitting with Al.

Distinct electronegativity difference between Al and H leads to strong polarization of the Al–H bond, resulting in negative charge enriching in H. Subsequently, H$^{\delta-}$ of Al–H species attacks the C$^{\delta+}$ of HCO$_3^-$, followed by the hydroxyl group of HCO$_3^-$ leaving. Finally, formate was obtained together with aluminum hydroxyl compound, which can dehydrate H$_2$O to form AlO(OH).

9.4 CONCLUSIONS

In summary, based on the experimental results and quantum chemistry calculation, the reduction of $NaHCO_3$ into formic acid by water splitting with Al under hydrothermal conditions is demonstrated to be a facile and highly efficient method for CO_2 reduction. The study can assist in developing an efficient and economical process for conversion of CO_2 to value-added chemicals. It provides us a promising approach to solve electronic waste metal utilization and CO_2 reduction simultaneously.

ACKNOWLEDGMENTS

The authors thank the financial support of the National Natural Science Foundation of China (No. 21277091), the State Key Program of National Natural Science Foundation of China (No. 21436007), Key Basic Research Projects of Science and Technology Commission of Shanghai (No. 14JC1403100), and China Postdoctoral Science Foundation (No. 2013 M541520).

REFERENCES

1. Sakakura, T., Choi, J. C., Yasuda, H. 2007. Transformation of carbon dioxide. *Chem. Rev.* 107:2365–87.
2. Wang, W., Wang, S. P., Ma, X. B., Gong, J. L. 2011. Recent advances in catalytic hydrogenation of carbon dioxide. *Chem. Soc. Rev.* 40:3703–27.
3. Yang, X. Y., Xiao, T. C., Edwards, P. P. 2011. The use of products from CO_2 photoreduction for improvement of hydrogen evolution in water splitting. *Int. J. Hydrogen Energy* 36:6546–52.
4. Handoko, A. D., Tang, J. W. 2013. Controllable proton and CO_2 photoreduction over Cu_2O with various morphologies. *Int. J. Hydrogen Energy* 38:13017–22.
5. Barton, E. E., Rampulla, D. M., Bocarsly, A. B. 2008. Selective solar-driven reduction of CO_2 to methanol using a catalyzed p-GaP based photoelectrochemical cell. *J. Am. Chem. Soc.* 130:6342–4.
6. Woolerton, T. W., Sheard, S., Pierce, E., Ragsdale, S. W., Armstrong, F. A. 2011. CO_2 photoreduction at enzyme-modified metal oxide nanoparticles. *Energy Environ. Sci.* 4:2393–9.
7. Hara, K., Kudo, A., Sakata, T. 1995. Electrochemical CO_2 reduction on a glassy carbon electrode under high pressure. *J. Electroanal Chem.* 391:141–7.
8. Galvez, M., Loutzenhiser, P., Hischier, I., Steinfeld, A. 2008. CO_2 splitting via two-step solar thermochemical cycles with Zn/ZnO and FeO/Fe_3O_4 redox reactions: Thermodynamic analysis. *Energy Fuels* 22:3544–50.
9. Chueh, W. C., Falter, C., Abbott, M., Scipio, D., Furler, P., Haile, S. M., Steinfeld, A. 2010. High-flux solar-driven thermochemical dissociation of CO_2 and H_2O using nonstoichiometric ceria. *Science* 330:1797–801.
10. Loutzenhiser, P. G., Steinfeld, A. 2011. Solar syngas production from CO_2 and H_2O in a two-step thermochemical cycle via Zn/ZnO redox reactions: Thermodynamic cycle analysis. *Int. J. Hydrogen Energy* 36:12141–7.
11. Wu, B., Gao, Y., Jin, F. M., Cao, J. L., Du, Y. X., Zhang, Y. L. 2009. Catalytic conversion of $NaHCO_3$ into formic acid in mild hydrothermal conditions for CO_2 utilization. *Catal. Today* 148:405–10.

12. Jin, F. M., Xu, Z., Liu, J. K., Jin, Y. J., Wang, L. Y., Zhong, H., Yao, G. D., Huo, Z. B. 2014. Highly efficient and autocatalytic H_2O dissociation for CO_2 reduction into formic acid with zinc. *Science Rep.* 4:4503.
13. Steinfeld, A., Kuhn, P., Reller, A., Palumbo, R., Murray, J., Tamaura, Y. 1998. Solar-processed metals as clean energy carriers and water-splitters. *Int. J. Hydrogen Energy* 23:767–74.
14. Steinfeld, A. 2002. Solar hydrogen production via a two-step water-splitting thermochemical cycle based on Zn/ZnO redox reactions. *Int. J. Hydrogen Energy* 27:611–19.
15. Dubreuil, J. F., Poliakoff, M. 2006. Organic reactions in high-temperature and supercritical water. *Pure Appl. Chem.* 78:1971–82.
16. Yamaguchi, A., Yamamoto, M., Takai, K., Ishii, T., Hashimoto, K., Nakamura, R. 2014. Electrochemical CO_2 reduction by Ni-containing iron sulfides: How is CO_2 electrochemically reduced at bisulfide-bearing deep-sea hydrothermal precipitates? *Electrochim. Acta* 141:311–18.
17. Alvarez-Barcia, S., Flores, J. R. 2009. The interaction of Al atoms with water molecules: A theoretical study. *J. Chem. Phys.* 131:174307.
18. McClean, R. E., Kauffman, J. W., Margrave, J. L. 1993. Kinetics of reaction Al + H_2O over an extended temperature. *J. Phys. Chem.* 97:9673–6.
19. Alexander, S., Nataliya, T., Alexander, S. 2011. Kinetics of Al + H_2O reaction: Theoretical study. *J. Phys. Chem. A* 115:4476–81.
20. Alvarez-Barcia, S., Jesus, R. F. 2011. A theoretical study of the dynamics of the Al + H_2O reaction in the gas-phase. *Chem. Phys.* 382:92–7.
21. Zeng, X., Hatakeyama, M., Ogata, K., Liu, J. K., Wang, Y. Q., Gao, Q. et al. 2014. New insights into highly efficient reduction of CO_2 to formic acid by using zinc under mild hydrothermal conditions: A joint experimental and theoretical study. *Phys. Chem. Chem. Phys.* 16:19836–40.
22. Jin, F. M., Zhou, Z. Y., Moriya, T., Kishida, H., Higashijima, H., Enomoto, H. 2005. Controlling hydrothermal reaction pathways to improve acetic acid production from carbohydrate biomass. *Environ. Sci. Technol.* 39:1893–902.
23. Wang, Y. Q., Jin, F. M., Zeng, X., Yao, G. D., Jing, Z. Z. 2013. A novel method for producing hydrogen from water with Fe enhanced by HS^- under mild hydrothermal conditions. *Int. J. Hydrogen Energy* 38:760–8.
24. Frisch, M. J., Trucks, G. W., Schlegel, H. B., Scuseria, G. E., Robb, M. A., Cheeseman, J. R. et al. 2009. Gaussian 09, revision B.1. Wallingford, CT: Gaussian Inc.
25. Sharipov, A., Titova, N., Starik, A. 2011. Kinetics of Al + H_2O reaction: Theoretical study. *J. Phys. Chem. A* 115:4476–81.
26. Hauge, R. H., Kauffman, J. W., Margrave, J. L. 1980. Infrared matrix-isolation studies of the interactions and reactions of Group 3A metal atoms with water. *J. Am. Chem. Soc.* 102:6005–11.
27. Joly, H. A., Howard, J. A., Tomietto, M., Tse, J. S. 1994. Characterization of the intermediates formed in the reaction of Al atoms with H_2O, H_2S and H_2Se by EPR spectroscopy. *J. Chem. Soc. Faraday Trans.* 90:3145–51.
28. Douglas, M. A., Hauge, R. H., Margrave, J. L. 1983. Matrix isolation studies by electronic spectroscopy of group IIIA metal water photochemistry. *J. Chem. Soc. Faraday Trans.* 179:1533–7.
29. Barcia, S. A., Flores, J. R. 2010. How H_2 can be formed by the interaction of Al atoms with a few water molecules: A theoretical study. *Chem. Phys.* 374:131–7.

10 Cu-Catalyzed Hydrothermal CO_2 Reduction with Zinc to Produce Methanol

Zhibao Huo, Dezhang Ren, Guodong Yao, Fangming Jin, and Mingbo Hu

CONTENTS

10.1 Introduction 141
10.2 Conversion of CO_2 to Methanol over Commercial Cu Powder 143
    10.2.1 Materials, Analysis, and Experimental Procedure 143
        10.2.1.1 General Information 143
        10.2.1.2 Product Analysis 143
        10.2.1.3 General Procedure for the Synthesis of Methanol from CO_2 ... 144
    10.2.2 Results and Discussion 144
        10.2.2.1 Effects of Reaction Parameters on the Yields of Methanol ... 144
        10.2.2.2 Possible Mechanism for the Production of Methanol over Commercial Cu Powder 147
10.3 Conversion of CO_2 into Methanol over the Cu Nanoparticle 149
    10.3.1 General Procedure for the Formation of Methanol 149
    10.3.2 Formation of Methanol over the Cu Nanoparticle 149
10.4 Conclusions and Prospects 150
References 150

10.1 INTRODUCTION

Methanol is one of the most important organic chemicals; it can be an alternative fuel for internal combustion engines. Compared with gasoline, methanol is more biodegradable and environmentally benign. Utilization of methanol as fuel directly or blending into gasoline is an efficient way to reduce the carbon footprint. Furthermore, transesterification or esterification of triglycerides or waste oil derived from biomass materials with methanol for production of biodiesel is another important application in the energy industry (Olah 2004, 2005). It is more meaningful in the context of worldwide carbon emissions reduction. Methanol is a versatile chemical: it has been used as an antifreeze; as a solvent; as a raw material for production of formaldehyde, flavors, dyes, medicines, and gunpowder; and as a common chemical feedstock for acetic acid, methyl *tert*-butyl ether, and chloromethane (Liu et al. 2003; Olah 2004, 2005; Ortelli et al. 2001). Synthesis of methanol from syngas (CO + H_2) is a commercial method, and some research has also been reported in previous works (Behrens et al. 2012; Lange 2001). However, because coal-based syngas is unsustainable, this method still has some drawbacks. Therefore, the development of a new method for the direct conversion of CO_2 into methanol is highly desirable. In recent years, synthesis of methanol from CO_2 is attracting interest because of its industrial importance, and much attention has been paid to the development of efficient methods for the synthesis of methanol, such as photochemical reduction (Anpo et al. 1998; Guan et al. 2003; Ikeue et al. 2001; Luo et al. 2011), electrochemical reduction (Hara et al. 1995, 1997; Li and Prentice 1997; Kaneco et al. 2000), and catalytic hydrogenation reduction (An et al. 2009; Deng et al. 1996; Fan et al. 1999; Guo et al. 2011; Jia et al. 2009; Jessop et al. 1996; Liaw and Chen 2001; Nitta et al. 1994; Roeda and Dollé 2006; Vesborg et al. 2009; Yang et al. 2010). Although these reactions have proven to be extremely efficient for the synthesis of methanol, most of them still have considerable drawbacks, including low yield, strict reaction conditions, and/or high cost. To avoid this, there is increasing demand for the development of a new method to effectively change CO_2 into methanol.

Over the past several decades, CO_2 hydrogenation to methanol is commonly used because of its abundance, low cost, nontoxicity, and high potential as a renewable source. It is well known that CO_2 has high thermodynamic stability and low reactivity; regardless of the method reported for methanol synthesis, the use of catalysts is essential. More importantly, the catalytic reduction of CO_2 into methanol has been proven technically feasible. Developing mild methods to catalytically activate CO_2 is a challenge for both academic and practical applications. Cu-based catalysts like Cu/ZnO or Cu/ZnO/Al_2O_3 have often been utilized for the methanol synthesis from syngas because the Cu element was active (Equation 10.1). It is, however, necessary to add metal oxides as adsorbed species (An et al. 2009; Deng et al. 1996; Fan et al. 1999; Guo et al. 2011; Jia et al. 2009; Jessop et al. 1996; Liaw and Chen 2001; Nitta et al. 1994; Roeda and Dollé 2006; Vesborg et al. 2009; Yang et al. 2010). Recently, hydrothermal treatment of converting CO_2 or biomass into value-added chemicals has received much attention and has become one of the most promising techniques because of its unique advantages (Berndt et al. 1996; He et al. 2010; Horita and Berndt 1999; Jin and Enomoto 2011; Shen et al. 2011; Tian et al. 2007; Voglesonger et al. 2001), such as fast reaction and green solvent. Previously, we and other groups have reported some research results in the hydrothermal conversion of CO_2 into value-added chemicals: (1) methane was formed via nickel-catalyzed conversion of CO_2 (Equation 10.2) (Takahashi et al. 2006) and (2) CO_2 was reduced to formic acid in the presence of Fe or Mn under hydrothermal conditions (Equation 10.3) (Duo et al. 2016; Jin et al. 2011; Lyu et al. 2014; Wu et al. 2009). Encouraged by these findings, it is possible that CO_2 hydrogenation could be converted into methanol using transition metal catalysts under hydrothermal conditions.

Here, we focused our attention on the conversion of CO_2 into methanol under hydrothermal conditions (Equation 10.4). In this process, water acts not only as an excellent solvent but also as a source of hydrogen generated by the reduction of metal reductants (Kruse and Dinjus 2007); hence, it would be advantageous to avoid the use of gaseous hydrogen, which is flammable, explosive, and not easy to operate during transportation. Among transition metals, we found that the Cu catalyst showed significant activity, in which Zn acts as an efficient reductant for methanol formation. Therefore, Zn and Cu, two commercially available and inexpensive metals, were chosen to be a reductant and a catalyst, respectively, under hydrothermal reactions. Detailed results are reported herein.

General methods for catalytic hydrogenation of CO_2 into CH_3OH.

$$CO_2 + H_2 \xrightarrow{\text{Cu-based catalysts}} CH_3OH \tag{10.1}$$

Our previous work. Hydrothermal conversion of CO_2 into chemicals.

$$CO_2 + H_2O \xrightarrow[\text{Hydrothermal condition}]{Cat.\ \text{Ni/Zn reductant}} CH_4 \tag{10.2}$$

$$CO_2 + H_2O \xrightarrow[\text{Hydrothermal condition}]{Cat.\ \text{Mn/Fe reductant}} HCOOH \tag{10.3}$$

This work: Hydrothermal conversion of CO_2 into CH_3OH.

$$CO_2 + H_2O \xrightarrow[\text{Hydrothermal condition}]{Cat.\ \text{Cu/Zn reductant}} CH_3OH \tag{10.4}$$

10.2 CONVERSION OF CO_2 TO METHANOL OVER COMMERCIAL CU POWDER

10.2.1 Materials, Analysis, and Experimental Procedure

10.2.1.1 General Information

$NaHCO_3$, methanol, Zn, HCl, and HCOOH were purchased from Sinopharm Chemical Reagent (China). Cu powder (200 mesh) was purchased from Shanghai Richjoint Co., Ltd. All chemicals were analytically pure. The reactor vessel was a piece of SUS 316 stainless steel batch tube with an inner volume of 42 mL that was connected to a Swagelok cap.

10.2.1.2 Product Analysis

After the reactions, the gas was collected and analyzed by gas chromatography equipped with a thermal conductivity detector (GC-TCD). The remaining reaction mixture was filtered, and then the precipitate was dried in an isothermal oven at 70°C for 24 h after washing with distilled water. The liquid sample was analyzed with high-performance liquid chromatography (HPLC) and GC-FID, and the precipitate was determined by an x-ray diffractometer (XRD).

10.2.1.3 General Procedure for the Synthesis of Methanol from CO_2

In this study, $NaHCO_3$ was used as the source of CO_2 to simplify the experiments. Experiments were conducted using a batch reactor system, and the schematic drawing can be found elsewhere (Jin et al. 2001).

A typical experimental procedure is as follows. First, the desired amount of different concentrations of the HCl solution was loaded into the reactor; then, Cu powder (catalyst), Zn powder (reductant), and $NaHCO_3$ (CO_2 source) were loaded into the reaction chamber in sequence to avoid a quick reaction with HCl, and the reactor was sealed. The reactor was put into the salt bath, which had been preheated to the desired temperature (250–350°C). During the reaction, the reactor was shaken while being kept horizontal to enhance the mixture and heat transfer. After the desired reaction time, the reactor was taken out of the salt bath and put into a cold water bath to quench the reaction. The reaction time was defined as the duration the reactor was kept in the salt bath. Filling ratio was defined as the ratio of the volume of the solution put into the reactor and the inner volume of the reactor.

10.2.2 Results and Discussion

10.2.2.1 Effects of Reaction Parameters on the Yields of Methanol

Initially, we screened the reaction conditions for the hydrothermal conversion of CO_2 into methanol. The reaction was carried out with 2.0 M NaOH at 350°C for 2 h in the presence of 70 mmol Cu and 60 mmol Zn in H_2O (water filling: 50%). It was found that no methanol formation was observed in the presence of NaOH, while the reaction of CO_2 proceeded to give the desired methanol in the presence of HCl. Therefore, we continued to optimize the initial reaction conditions for the formation of methanol from CO_2.

First, we carried out the experiments with 2.0 M HCl at 350°C for 2 h in the presence of 70 mmol Cu and 60 mmol Zn in H_2O (water filling: 50%) to examine the effect of Cu and Zn in hydrothermal reaction. After the hydrothermal reaction, solid residual was analyzed by XRD. As illustrated in Figure 10.1, Cu showed no apparent change and still existed in solid samples while Zn was converted into ZnO. Some unknown phases in the precipitates after hydrothermal reaction were also observed. It is just speculative for some unknown phases to be CuO or Cu_2O according to limited experimental data. From the results, it is clarified that Zn acts as a reductant, and Cu has catalytic activity for reduction of CO_2 into methanol as a catalyst.

The effect of Cu amount on the formation of methanol from CO_2 was observed at 350°C for 2 h with the different ratios of $Zn/NaHCO_3$. The results are shown in Figure 10.2. The profile of the reaction of CO_2 monitored by HPLC indicated that the yield of methanol increased by changing the Cu amount from 20 to 70 mmol in all cases. Furthermore, the $Zn/NaHCO_3$ ratio of 1:1 gave a higher yield of methanol as compared to the $Zn/NaHCO_3$ ratio of 1.5:1. A small increase in the yield of methanol was observed in the presence of Cu amount from 20 to 40 mmol. Interestingly, the Cu catalyst showed significantly catalytic activity for methanol production and the yields of methanol increased as the Cu amount increased from 40 to 70 mmol to reach the maximum yields of 6.5% and 10.4%, respectively, for methanol. The highest yield

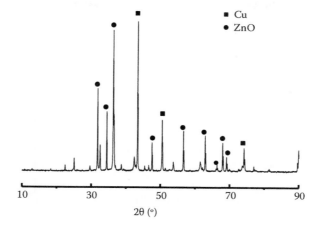

FIGURE 10.1 XRD pattern of precipitates after hydrothermal reaction (*T*: 350°C; time: 2 h; Cu: 70 mmol; Zn: 60 mmol; NaHCO$_3$: 40 mmol; HCl: 2.0 M; water filling: 50%).

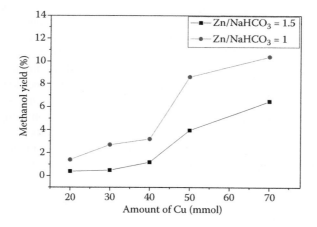

FIGURE 10.2 Effect of amount of Cu on the yields of methanol (*T*: 350°C; time: 2 h; Zn: 60 mmol; NaHCO$_3$: 40 [or 60] mmol; HCl: 2.0 M; water filling: 50%).

of methanol was obtained at 2 h when Cu (70 mmol) was added in the case of a Zn/NaHCO$_3$ ratio of 1:1.

To examine the effect of concentration of HCl on yield of methanol, experiments were carried out by varying HCl concentrations from 0 to 3 M at 350°C for 2 h in the presence of 50 mmol Cu and 60 mmol Zn. As shown in Figure 10.3, the yield of methanol remarkably increased as the concentration of HCl increased from 0 to 2.0 M. Between HCl concentrations of 0.5 and 1.2 M, the yield of methanol was almost constant. The yield reached a maximum value of 4.2% for methanol at a HCl concentration of 2.0 M. The yield was decreased drastically with the further increase of HCl concentration after attaining a peak level at a concentration of 2.0 M. It is probably

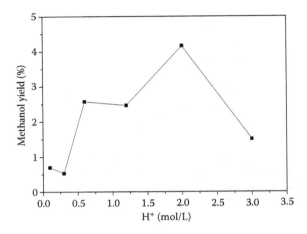

FIGURE 10.3 Effect of concentration of HCl on the yield of methanol (T: 350°C; time: 2 h; Cu: 50 mmol; Zn: 60 mmol; NaHCO$_3$: 40 mmol; water filling: 50%).

because of the catalyst deactivation caused by chlorine. It has been reported that a trace of chlorine can lead to a great loss of the activity of the Cu-based catalyst (Liu et al. 2003). The optimum HCl concentration was 2.0 M to achieve the maximum yield of methanol from CO$_2$. The present study indicated that HCl concentration had a greater effect on the yield of methanol under hydrothermal conditions.

It is known that CO$_2$ has high thermodynamic stability and low reactivity. Therefore, a high reaction temperature is favorable for the conversion of CO$_2$. Experiments were performed to examine the effect of reaction temperature on the yields of methanol from CO$_2$ at different reaction temperatures from 250°C to 350°C for 2 h. The results in Figure 10.4 show that the yields of methanol remarkably increased as the temperature increased from 250°C to 350°C. The reaction

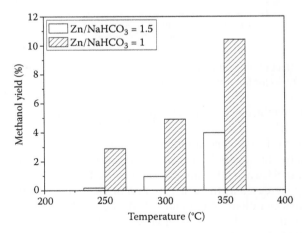

FIGURE 10.4 Effect of temperature on the yield of methanol (time: 2 h; Cu: 50 mmol; Zn: 60 mmol; NaHCO$_3$: 40 [or 60] mmol; water filling: 50%; HCl: 2.0 M).

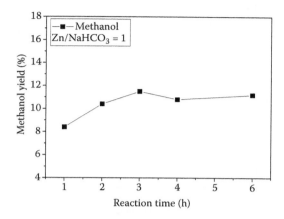

FIGURE 10.5 Effect of reaction time on the yield of methanol (T: 350°C; Cu: 70 mmol; Zn: 60 mmol; NaHCO$_3$: 40 [or 60] mmol; water filling: 50%; HCl: 2.0 M).

temperature was 350°C to achieve the maximum value of 10.4% methanol from CO$_2$. In addition, the Zn/NaHCO$_3$ ratio of 1:1 gave a higher yield of methanol as compared to the Zn/NaHCO$_3$ ratio of 1.5:1. It means that a large amount of NaHCO$_3$ can improve the yield of methanol under hydrothermal conditions.

Experiments were performed by changing the reaction time from 1 to 6 h with 70 mmol Cu, 60 mmol Zn, and 2.0 M HCl at 350°C to examine the effect of reaction time. Variation of methanol yields is shown in Figure 10.5. The time profile of the reaction of CO$_2$ indicated that the yields increased as reaction time increased in the first 3 h. The highest yield of methanol, 11.4%, was obtained when the reaction time was 3 h. The yield of methanol had a small change and decreased a little from 3 to 4 h. The formation of methanol reached a plateau after 4 h and the yield did not increase significantly even at a prolonged reaction time. Therefore, a longer reaction time has little influence on the yield of methanol.

The effect of filling rate on the yield of methanol was also investigated at different water filling values from 35% to 50% for 3 h when the Zn/NaHCO$_3$ ratio is 1:1 and 1.5:1. As shown in Figure 10.6, the yields of methanol increased as the water filling increased from 35% to 50% in all cases. In the case of the Zn/NaHCO$_3$ ratio of 1.5:1, the yield of methanol increased from 0.6% (water filling: 35%) to 3.96% (water filling: 50%). The optimum water filling was 50% to achieve the maximum value of 11.4% methanol from CO$_2$. The Zn/NaHCO$_3$ ratio of 1:1 showed a higher yield than the Zn/NaHCO$_3$ ratio of 1.5:1.

10.2.2.2 Possible Mechanism for the Production of Methanol over Commercial Cu Powder

Following the reaction, the gas samples were collected for analysis by using GC-TCD. The experimental results are shown in Table 10.1. Only H$_2$ and CO$_2$ were detected in gas samples, and almost no CO and CH$_4$ was detected, which indicated that the reactions concerning CO and CH$_4$ did not occur. Compared with the two experiments, CO$_2$ was nearly used up in experiment 1.

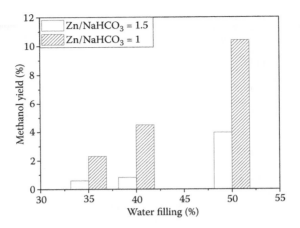

FIGURE 10.6 Effect of water filling on the yield of methanol (T: 350°C; Cu: 70 mmol; Zn: 60 mmol; $NaHCO_3$: 40 [or 60] mmol; HCl: 2.0 M).

TABLE 10.1
The Percentage of H_2 and CO_2 for Gas Samples after the Reactions

	Experiment 1[a]	Experiment 2[b]
H_2	0.98	0.73
CO_2	0.031	0.253

[a] With reductant and catalyst.
[b] Without reductant and catalyst.

The mechanism for methanol synthesis from CO_2/H_2 has been proposed by a number of researchers. For example, several research groups have investigated the $Cu/ZrO_2/SiO_2$ catalysts by in situ infrared spectroscopy and suggested that Cu supported on zirconia is exceptionally active for methanol synthesis from CO_2 (Clarke and Bell 1995; Fisher and Bell 1997). Therefore, the addition of zirconia to Cu/SiO_2 progressively enhances the activity of the catalyst for methanol synthesis from CO_2. Liu reported that ZnO can act synergistically with Cu to catalyze the synthesis of methanol, and ZnO was a good hydrogenation catalyst that activates hydrogen by heterogeneous splitting (Liu et al. 2003). According to these findings, the proposed path of reduction of CO_2 into methanol with Cu in the presence of Zn under hydrothermal conditions is illustrated in Figure 10.7. Initially, CO_2 is adsorbed on ZnO and subsequently undergoes stepwise hydrogenation to methoxide species, with atomic hydrogen being supplied by spillover from Cu. The final step is the hydrolysis of methoxide groups on ZnO to obtain methanol along with H_2O produced as a co-product of methanol formation.

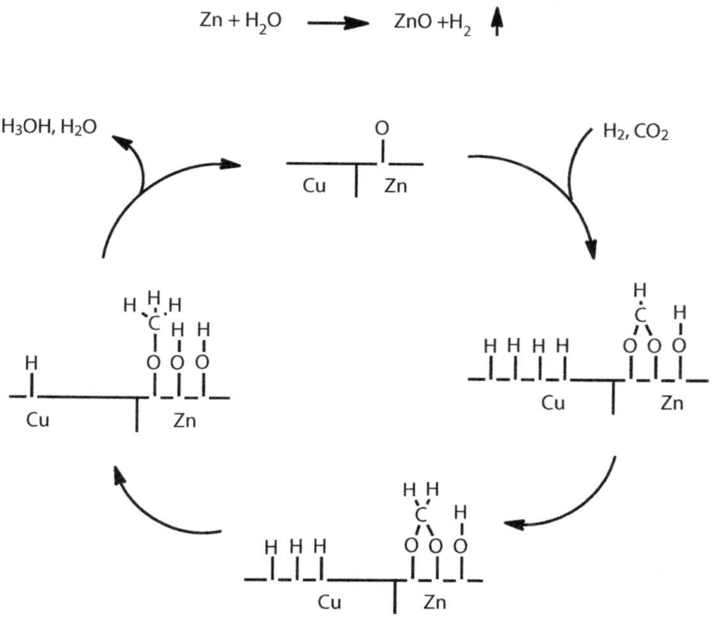

$$Zn + H_2O \longrightarrow ZnO + H_2 \uparrow$$

FIGURE 10.7 Plausible mechanism for the formation of methanol from CO_2 over Cu.

10.3 CONVERSION OF CO_2 INTO METHANOL OVER THE CU NANOPARTICLE

10.3.1 General Procedure for the Formation of Methanol

A typical experimental procedure is as follows. First, the desired amount of HCl solution (2 mol/L) was loaded into the reactor and then the Cu nanoparticle (50 nm), Zn powder, and $NaHCO_3$ (60 mmol) were loaded into the reaction chamber in sequence to avoid a quick reaction with HCl. The reactor was sealed and then was put into the salt bath, which had been preheated to the desired temperature. During the reaction, the reactor was shaken while being kept horizontal to enhance the mixture and heat transfer. After the desired reaction time, the reactor was taken out of the salt bath and put into a cold water bath to quench the reaction. The reaction time was defined as the duration the reactor was kept in the salt bath.

10.3.2 Formation of Methanol over the Cu Nanoparticle

We investigated the catalytic efficiency of the Cu nanoparticle (50 nm) compared with the Cu powder (200 mesh) for methanol production. The specific surface area of the Cu nanoparticle (15 m^2/g) is three times larger than that of the Cu powder. The reaction was carried out in the presence of 60 mmol Zn and 60 mmol Cu nanoparticle with 50% water filling at 300°C for 2 h and gave a desired methanol yield of 10.8%. Although the Cu nanoparticle has a higher specific area, the yield of methanol was almost identical to the commercial one. The result indicated that the Cu nanoparticle was also an efficient catalyst for CO_2 conversion.

10.4 CONCLUSIONS AND PROSPECTS

In summary, hydrothermal conversion of CO_2 into methanol using Zn as a reductant and Cu as a catalyst has been proven to be an efficient method. Cu powder exhibits a certain catalytic activity and the highest yield of methanol is 11.4%. Although the efficiency of this reaction was not very high, the catalytic conversion of CO_2 into methanol in high-temperature water is possibly realized and the process is also green and environmentally friendly. Currently, we are still exploring new Cu-based catalysts to improve methanol yield. The present study is helpful not only for future practical applications in controlling the global greenhouse effect but also for searching for useful alternative fuels or carbon resources.

REFERENCES

An, X., Y. Zuo, Q. Zhang, and J. Wang. 2009. Methanol synthesis from CO_2 hydrogenation with a Cu/Zn/Al/Zr fibrous catalyst. *Chinese J. Chem. Eng.* 17 (1):88–94.

Anpo, M., H. Yamashita, K. Ikeue, Y. Fujii, S. G. Zhang, Y. Ichihashi, D. R. Park, Y. Suzuki, K. Koyano, and T. Tatsumi. 1998. Photocatalytic reduction of CO_2 with H_2O on Ti-MCM-41 and Ti-MCM-48 mesoporous zeolite catalysts. *Catal. Today* 44 (1–4):327–332.

Behrens, M., F. Studt, I. Kasatkin, S. Kuhl, M. Havecker, F. Abild-Pedersen, S. Zander, F. Girgsdies, P. Kurr, B. L. Kniep, M. Tovar, R. W. Fischer, J. K. Norskov, and R. Schlogl. 2012. The active site of methanol synthesis over $Cu/ZnO/Al_2O_3$ industrial catalysts. *Science* 336 (6083):893–897.

Berndt, M. E., D. E. Allen, and W. E. Seyfried. 1996. Reduction of CO_2 during serpentinization of olivine at 300°C and 500 bar. *Geology* 24 (4):351–354.

Clarke, D. B., and A. T. Bell. 1995. An infrared study of methanol synthesis from CO_2 on clean and potassium-promoted Cu/SiO_2. *J. Catal.* 154 (2):314–328.

Deng, J., Q. Sun, Y. Zhang, S. Chen, and D. Wu. 1996. A novel process for preparation of a $Cu/ZnO/Al_2O_3$ ultrafine catalyst for methanol synthesis from $CO_2 + H_2$: Comparison of various preparation methods. *Appl. Catal. A: Gen.* 139 (1–2):75–85.

Duo, J., F. Jin, Y. Wang, H. Zhong, L. Lyu, G. Yao, and Z. Huo. 2016. $NaHCO_3$-enhanced hydrogen production from water with Fe and in situ highly efficient and autocatalytic $NaHCO_3$ reduction into formic acid. *Chem Commun (Camb)* 52 (16):3316–3319.

Fan, L., Y. Sakaiya, and K. Fujimoto. 1999. Low-temperature methanol synthesis from carbon dioxide and hydrogen via formic ester. *Appl. Catal. A: Gen.* 180 (1–2):L11–L13.

Fisher, I. A., and A. T. Bell. 1997. *In-situ* infrared study of methanol synthesis from H_2/CO_2 over Cu/SiO_2 and $Cu/ZrO_2/SiO_2$. *J. Catal.* 172 (1):222–237.

Guan, G., T. Kida, and A. Yoshida. 2003. Reduction of carbon dioxide with water under concentrated sunlight using photocatalyst combined with Fe-based catalyst. *Appl. Catal. B: Environ.* 41 (4):387–396.

Guo, X., D. Mao, G. Lu, S. Wang, and G. Wu. 2011. CO_2 hydrogenation to methanol over $Cu/ZnO/ZrO_2$ catalysts prepared via a route of solid-state reaction. *Catal. Commun.* 12 (12):1095–1098.

Hara, K., A. Kudo, and T. Sakata. 1995. Electrochemical reduction of carbon dioxide under high pressure on various electrodes in an aqueous electrolyte. *J. Electroanal. Chem.* 391 (1–2):141–147.

Hara, K., A. Tsuneto, A. Kudo, and T. Sakata. 1997. Change in the product selectivity for the electrochemical CO_2 reduction by adsorption of sulfide ion on metal electrodes. *J. Electroanal. Chem.* 434 (1–2):239–243.

He, C., G. Tian, Z. Liu, and S. Feng. 2010. A mild hydrothermal route to fix carbon dioxide to simple carboxylic acids. *Org. Lett.* 12 (4):649–651.

Horita, J., and M. E. Berndt. 1999. Abiogenic methane formation and isotopic fractionation under hydrothermal conditions. *Science* 285 (5430):1055–1057.

Ikeue, K., H. Yamashita, M. Anpo, and T. Takewaki. 2001. Photocatalytic reduction of CO_2 with H_2O on Ti–β zeolite photocatalysts: Effect of the hydrophobic and hydrophilic properties. *J. Phys. Chem. B* 105 (35):8350–8355.

Jessop, P. G., Y. Hsiao, T. Ikariya, and R. Noyori. 1996. Homogeneous catalysis in supercritical fluids: Hydrogenation of supercritical carbon dioxide to formic acid, alkyl formates, and formamides. *J. Am. Chem. Soc.* 118 (2):344–355.

Jia, L., J. Gao, W. Fang, and Q. Li. 2009. Carbon dioxide hydrogenation to methanol over the pre-reduced $LaCr_{0.5}Cu_{0.5}O_3$ catalyst. *Catal. Commun.* 10 (15):2000–2003.

Jin, F., and H. Enomoto. 2011. Rapid and highly selective conversion of biomass into value-added products in hydrothermal conditions: Chemistry of acid/base-catalysed and oxidation reactions. *Energy Environ. Sci.* 4 (2):382–397.

Jin, F., Y. Gao, Y. Jin, Y. Zhang, J. Cao, Z. Wei, and R. L. Smith Jr. 2011. High-yield reduction of carbon dioxide into formic acid by zero-valent metal/metal oxide redox cycles. *Energy Environ. Sci.* 4 (3):881–884.

Jin, F., A. Kishita, T. Moriya, and H. Enomoto. 2001. Kinetics of oxidation of food wastes with H_2O_2 in supercritical water. *J. Supercrit. Fluid.* 19 (3):251–262.

Kaneco, S., K. Iiba, K. Ohta, and T. Mizuno. 2000. Reduction of carbon dioxide to petrochemical intermediates. *Energy Sources* 22 (2):127–135.

Kruse, A., and E. Dinjus. 2007. Hot compressed water as reaction medium and reactant. *J. Supercrit. Fluid.* 39 (3):362–380.

Lange, J.-P. 2001. Methanol synthesis: A short review of technology improvements. *Catal. Today* 64 (1–2):3–8.

Li, J., and G. Prentice. 1997. Electrochemical synthesis of methanol from CO_2 in high-pressure electrolyte. *J. Electrochem. Soc.* 144 (12):4284–4288.

Liaw, B. J., and Y. Z. Chen. 2001. Liquid-phase synthesis of methanol from CO_2/H_2 over ultrafine CuB catalysts. *Appl. Catal. A: Gen.* 206 (2):245–256.

Liu, X.-M., G. Q. Lu, Z.-F. Yan, and J. Beltramini. 2003. Recent advances in catalysts for methanol synthesis via hydrogenation of CO and CO_2. *Ind. Eng. Chem. Res.* 42 (25):6518–6530.

Luo, D., Y. Bi, W. Kan, N. Zhang, and S. Hong. 2011. Copper and cerium co-doped titanium dioxide on catalytic photo reduction of carbon dioxide with water: Experimental and theoretical studies. *J. Mol. Struct.* 994 (1–3):325–331.

Lyu, L., X. Zeng, J. Yun, F. Wei, and F. Jin. 2014. No catalyst addition and highly efficient dissociation of H_2O for the reduction of CO_2 to formic acid with Mn. *Environ. Sci. Technol.* 48 (10):6003–6009.

Nitta, Y., O. Suwata, Y. Ikeda, Y. Okamoto, and T. Imanaka. 1994. Copper-zirconia catalysts for methanol synthesis from carbon dioxide: Effect of ZnO addition to $Cu-ZrO_2$ catalysts. *Catal. Lett.* 26 (3–4):345–354.

Olah, G. A. 2004. After oil and gas: Methanol economy. *Catal. Lett.* 93 (1/2):1–2.

Olah, G. A. 2005. Beyond oil and gas: The methanol economy. *Angew. Chem. Int. Ed. Engl.* 44 (18):2636–2639.

Ortelli, E. E., J. Wambach, and A. Wokaun. 2001. Methanol synthesis reactions over a CuZr based catalyst investigated using periodic variations of reactant concentrations. *Appl. Catal. A: Gen.* 216 (1–2):227–241.

Roeda, D., and F. Dollé. 2006. [¹¹C]Methanol production by a fast and mild aqueous-phase reduction of [¹¹C]formic acid with samarium diiodide. *J. Labelled Compd Rad.* 49 (3):295–304.

Shen, Z., Y. Zhang, and F. Jin. 2011. From $NaHCO_3$ into formate and from isopropanol into acetone: Hydrogen-transfer reduction of $NaHCO_3$ with isopropanol in high-temperature water. *Green Chem.* 13 (4):820–823.

Takahashi, H., L. H. Liu, Y. Yashiro, K. Ioku, G. Bignall, N. Yamasaki, and T. Kori. 2006. CO_2 reduction using hydrothermal method for the selective formation of organic compounds. *J. Mater. Sci.* 41 (5):1585–1589.

Tian, G., H. Yuan, Y. Mu, C. He, and S. Feng. 2007. Hydrothermal reactions from sodium hydrogen carbonate to phenol. *Org. Lett.* 9 (10):2019–2021.

Vesborg, P. C. K., I. Chorkendorff, I. Knudsen, O. Balmes, J. Nerlov, A. M. Molenbroek, B. S. Clausen, and S. Helveg. 2009. Transient behavior of Cu/ZnO-based methanol synthesis catalysts. *J. Catal.* 262 (1):65–72.

Voglesonger, K. M., J. R. Holloway, E. E. Dunn, P. J. Dalla-Betta, and P. A. O'Day. 2001. Experimental abiotic synthesis of methanol in seafloor hydrothermal systems during diking events. *Chem. Geol.* 180 (1–4):129–139.

Wu, B., Y. Gao, F. Jin, J. Cao, Y. Du, and Y. Zhang. 2009. Catalytic conversion of $NaHCO_3$ into formic acid in mild hydrothermal conditions for CO_2 utilization. *Catal. Today* 148 (3–4):405–410.

Yang, Y., J. Evans, J. A. Rodriguez, M. G. White, and P. Liu. 2010. Fundamental studies of methanol synthesis from CO_2 hydrogenation on Cu(111), Cu clusters, and Cu/ZnO(0001). *Phys. Chem. Chem. Phys.* 12 (33):9909–9917.

11 Hydrothermal Reduction of CO_2 with Glycerine

Zheng Shen, Minyan Gu, Meng Xia, Wei Zhang, Yalei Zhang, and Fangming Jin

CONTENTS

11.1 Introduction 154
11.2 Experimental Section 155
    11.2.1 Materials 155
    11.2.2 Experimental Procedure 155
    11.2.3 Product Analysis 156
    11.2.4 NMR Analysis 156
    11.2.5 LC-MS Analysis 156
    11.2.6 HPLC Analysis 157
    11.2.7 GC Analysis 157
11.3 The Alcohol-Mediated Reduction of CO_2 and $NaHCO_3$ into Formate: A Hydrogen Transfer Reduction of $NaHCO_3$ with Glycerine under Alkaline Hydrothermal Conditions 157
    11.3.1 A Set of Comparative Experiments with or without Dry Ice (CO_2 Source) by Glycerine 157
    11.3.2 The Effects of $NaHCO_3$ Quantity and NaOH Concentration 159
    11.3.3 The Effects of Reaction Temperature and Time 160
    11.3.4 The Catalysis Effect of Reactor Materials 160
    11.3.5 The Reaction Pathway on the Hydrogen-Transfer Reduction of $NaHCO_3$ with Glycerine 161
11.4 From $NaHCO_3$ into Formate and from Isopropanol into Acetone: Hydrogen-Transfer Reduction of $NaHCO_3$ with Isopropanol in HTW 163
    11.4.1 Reaction Products Analysis from Isopropanol and $NaHCO_3$ 163
    11.4.2 The Effect of the Reaction Temperature 163
    11.4.3 The Effect of the Quantity of $NaHCO_3$ 164
    11.4.4 The Effect of the NaOH Concentration 165
    11.4.5 The Effect of the Reaction Time 166
    11.4.6 The Catalysis Effect of Reactor Materials 167
    11.4.7 The Reaction Pathway on the Hydrogen-Transfer Reduction of $NaHCO_3$ with Isopropanol 167
11.5 The Mechanism for Production of Abiogenic Formate from CO_2 and Lactate from Glycerine: Uncatalyzed Transfer Hydrogenation of CO_2 with Glycerine under Alkaline Hydrothermal Conditions 168
    11.5.1 Formation of Abiogenic Formate from CO_2 169

11.5.2 Effect of D_2O Solvent 171
11.5.3 Effect of Reactor Materials.. 173
11.5.4 Catalysis of H_2O Molecules.. 173
11.5.5 Proposition of Reaction Mechanism... 174
11.5.6 Testing of Reaction Mechanism 175
11.5.7 Verification of Alkaline Role... 176
11.5.8 Alkaline Hydrothermal Conversion Routes of Lactate
       and Intermediates 177
11.6 Conclusions.. 179
References.. 180

11.1 INTRODUCTION

If carbon dioxide, a stable carbon end product of metabolism and other combustions, could be effectively reduced into useful organic products, it would provide an abundant, low-value carbon source to be recycled and reused [1–4]. Significant efforts have been devoted to explore technologies for CO_2 transformation in which high free energy content substances (such as hydrogen, unsaturated compounds, small-membered ring compounds, and organometallics) are required to compensate for the high thermodynamic stability and low energy level of CO_2 [5–9]. In 1935, Farlow and Adkins reported the first direct hydrogen-transfer reduction of carbon dioxide to formic acid using H_2 as a reductant and Raney nickel as a catalyst [10]. Since then, the hydrogenation of carbon dioxide to form formic acid or formate salts has been widely investigated using homogeneous or heterogeneous metal catalysts [7,11–15]. Formic acid is an important chemical feedstock and is used as a synthetic precursor and a commercial product in the leather, agriculture, and dye industries. In addition, formic acid could be used as a raw material for hydrogen production and has the potential to power fuel cells to generate electricity and run automobiles [16–19]. However, the production of CO_2-derived formic acid is not widely used in industrial chemical processes because the reductant, H_2, is currently produced from reactions of crude oil-derived methane with water. Therefore, the development of an alternative substance that can reduce CO_2 is highly desirable.

Hydrothermal processes have been attracting increasing attention for use in organic chemical reactions because high-temperature water (HTW) is an environmentally benign solvent compared to organic solvents and has remarkable properties as a reaction medium [20–22]. For example, HTW has a lower dielectric constant, fewer and weaker hydrogen bonds, and a higher isothermal compressibility than ambient liquid water [20]. The solubility of most gases in liquid water initially decreases as the temperature is increased above ambient temperature, but a minimum solubility is soon reached, after which the gas solubility increases. For example, the minimal solubility of CO_2 occurs at around 150°C [22]. Moreover, many organic reactions under ambient conditions only proceed in the presence of acidic/basic or metallic catalysts; however, these reactions can occur in HTW in the absence of an added catalyst [23–26].

There is increasing interest in performing CO_2 reductions in HTW [27–33] because in this medium, H_2 can be produced from metal or alcohol compounds. Horita and Berndt reported that CO_2 is converted to CH_4 by H_2, which is formed under hydrothermal conditions (≤400°C, ≤100 MPa) in a process catalyzed by a hydrothermally formed nickel–iron alloy. In this system, H_2 is produced during the conversion of olivine into serpentine and magnetite [27]. It has also been found that formic acid can be hydrothermally produced from CO_2 with Fe powder and/or Ni powder [28–31]. He et al. have used iron nanoparticles not only as reducing agents but also as catalysts to transform CO_2 into formic acid and acetic acid [31]. For these metal-reducing agents, however, the reactivity in HTW is low, and the yield of formic acid was less than 3%. Our recent research showed that in HTW, $NaHCO_3$ is reduced to formate when isopropanol or glycerine is used as a reducing agent [32]. Accordingly, it is likely that industrial applications of CO_2 reduction would use abundant alcohol compounds as reducing materials.

11.2 EXPERIMENTAL SECTION

11.2.1 Materials

Glycerine (99%) and isopropanol (99%) were used as the test material. NaOH (96%) and $NaHCO_3$ (99%) were used as alkaline catalysts. Dry ice (99%) and $NaHCO_3$ were used as CO_2 sources. Glycerine, NaOH, $NaHCO_3$, dry ice, and formic acid (99.9%) were obtained from Sinopharm Chemical Reagent Co., Ltd, China. Other chemicals, such as $NaH^{13}CO_3$ (99% ^{13}C), D_2O (99.9% D), acetol (99%), pyruvaldehyde (99%), and L+lactate (99%), were supplied by Sigma-Aldrich (Shanghai) Trading Co., Ltd, China. In this study, $NaHCO_3$ was used as a CO_2 resource to simplify operations and to allow for an accurate quantification of CO_2 [30].

NaOD used were prepared by dissolving the solid base in deuterium oxide, followed by drying in a rotary evaporator and then repeating the process three times. Although the resulting NaOD solid contained some residual ^1H, the amount of ^1H introduced into the reaction mixture was negligible as only a small quantity of base relative to the deuterium oxide solvent was used in each reaction. $NaDCO_3$ used in this investigation were prepared by aerating excess CO_2 in Na_2CO_3 solution of deuterium oxide for 2 h and then drying in a rotary evaporator.

The schematic drawing of the experimental setup can be found elsewhere [30,34]. Most experiments were performed in a batch-type reactor made of stainless steel 316 tubing (3/8 inch diameter, 1-mm wall thickness, 120-mm length) with two end fittings, providing an inner volume of 5.7 mL [33]. The reactor can collect gas by a nozzle and high-pressure valve. The reaction temperature was controlled by a salt bath. In a few cases, a batch reactor with a Teflon inner wall having an inner volume of 20 mL was used in order to investigate the effect of the reactor materials.

11.2.2 Experimental Procedure

The typical procedure of hydrothermal reactions is described as follows: A 4-mL water mixture with 0.33 M glycerine or 0.25 M isopropanol, 0–2.5 M NaOH (or NaOD), and 0–0.44 g of dry ice (CO_2) or 0–1.76 g of $NaHCO_3$ (or $NaDCO_3$) was added to the batch reactor, and then the reactor was put into a salt bath preheated to a desired temperature. In the salt bath, the reactor was shaken while being kept horizontally, to mix well and enhance heat transfer. After a desired reaction time, the reactor was removed from the salt bath and put into a cold-water bath to quench the reaction. The reaction time was defined as the period during which the reactor was kept in the salt bath. The real reaction time is shorter than the apparent reaction time, because the heat-up time to raise the temperature of the reaction media from 20°C to 300°C was approximately 15 s. The temperature of the salt bath was taken as the reaction temperature. After cooling, samples of the liquid phase and gas phase in the reactor were collected for analysis. In all experiments, we fixed the temperature at 300°C and water filling rate at 70%. Thus, the reaction pressure was approximately 9 MPa, which could be estimated from the water saturation pressure at 300°C.

In order to investigate the effect of the reactor materials on the hydrothermal reactions, a batch reactor with a Teflon inner wall having an inner volume of 20 mL was used, which had been described elsewhere [26]. The typical reaction procedure by using this reactor is as follows. A 14-mL mixture with 0.33 M glycerine, 1.40 g of NaOH, and 1.54 g of CO_2 was put into the reactor. After being sealed, the reactor was placed in an electric furnace that had been preheated to 300°C. After the desired reaction time, the reactor was removed from the electric furnace for cooling at room temperature (25°C). Then, liquid samples were collected for HPLC analysis.

11.2.3 Product Analysis

After the reactions, the liquid samples were collected for ^{1}H- and ^{2}H-NMR, ^{13}C-NMR, LC-MS, and HPLC analyses, and gas samples were collected for GC analysis. Peak identification was accomplished by comparing the sample peak retention time with those of standard solutions of pure compounds. All quantitative data reported in this study were the average values of the analytical results of at least three samples with the relative errors always less than 10% for all experiments.

11.2.4 NMR Analysis

^{1}H-, ^{2}H-, and ^{13}C-NMRs were performed using an NMR spectrometer (DMX 500, 500 MHz). In order to reduce the signal interference from the large number of solvent water, 40-µL collected liquid samples were put into 5 mm I.D. NMR tubes and then were dealt with through the freeze-drying process. After freeze-drying, 0.5 mL of D_2O was added into the sample NMR tubes for ^{1}H-NMR analyses or into 0.5 mL of H_2O for ^{2}H-NMR analyses.

11.2.5 LC-MS Analysis

Liquid chromatography was performed on a Shimadzu LC-10AD HPLC system consisting of an autosampler (SILHTc). The HPLC was coupled to a Shimadzu LCMS-2010A single quadrupole mass spectrometer with an electrospray ionization (ESI) interface. Data acquisition and processing were accomplished using Shimadzu LC-MS solution software for LC-MS-2010 high-performance liquid chromatography/mass spectrometer. LC-MS conditions are shown as follows: column, RSpak KC-811 (300 × 8.0 mm I.D., Shodex Packed Corporation, Japan); mobile phase, 1 mM $HClO_4$; flow rate, 0.25 mL/min; detection, UV 210 nm; column temperature, 50°C; MS, negative mode (scan range, m/z 30–120); ionization, ESI.

11.2.6 HPLC Analysis

The liquid samples were filtered through a 0.45-mm filter and then were adjusted with sulfuric acid until the pH values of the solution reached 2–3. After that, the liquid samples were analyzed by HPLC. HPLC analysis was performed using an Agilent 1200 HPLC system equipped with a tunable absorbance detector (UV detector) and a differential refractometer (RI detector). During the HPLC analysis, two columns (RSpak KC811) were used in series, and the solvent used was 1 mM $HClO_4$ at a flow rate of 1.0 mL/min.

11.2.7 GC Analysis

The gas sample was analyzed by an Agilent 7890 GC with a Porapak Q column or a POLA column.

11.3 THE ALCOHOL-MEDIATED REDUCTION OF CO_2 AND $NaHCO_3$ INTO FORMATE: A HYDROGEN TRANSFER REDUCTION OF $NaHCO_3$ WITH GLYCERINE UNDER ALKALINE HYDROTHERMAL CONDITIONS

Glycerine, namely, 1,2,3-propanetriol, is a potentially important biorefinery feedstock and is a by-product of biodiesel production. It is produced during the transesterification of vegetable oils or animal fats [35,36]. In our previous study [37–39], we reported that glycerine is efficiently converted to lactic acid under alkaline hydrothermal conditions. In addition, H_2 is produced in almost the same yield as lactic acid. Lactic acid is receiving attention as a material for producing biodegradable lactic acid polymers [40]. We now report that $NaHCO_3$ (as a source of CO_2) is converted to a formate salt when glycerine is used as a reducing agent in HTW. Glycerine is converted to lactate under these reaction conditions. The effect of reaction conditions was also investigated in detail, for example, $NaHCO_3$ quantity, NaOH concentration, reaction temperature, and time.

11.3.1 A Set of Comparative Experiments with or without Dry Ice (CO_2 Source) by Glycerine

First, a set of comparative experiments with glycerine and 0.25 M NaOH with or without dry ice (CO_2 source) were carried out at 300°C to determine whether CO_2 could be reduced to formate under these conditions. In our previous studies, 300°C was found to be an optimal temperature for hydrothermal reactions with glycerine [32,37–39]. HPLC traces of the reaction mixture are shown in Figure 11.1. When one compares Figure 11.1a to Figure 11.1b, it is clear that a large amount of formic acid was formed. The product yields and the remaining amounts of glycerine are detailed in Table 11.1. The product yield is defined as the percentage of the mole of products relative to the initial moles of the reducing agent glycerine. As shown in columns a and b of Table 11.1, the addition of dry ice resulted in a rapid increase in the formate yield and a relative decrease in H_2 yield, but no substantial change is observed in the

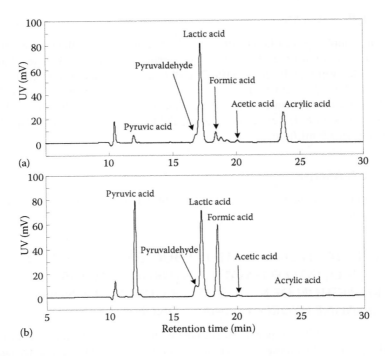

FIGURE 11.1 HPLC chromatograms of solutions after reaction of glycerine without or with dry ice (temperature: 300°C; reaction time: 60 min; glycerine: 0.33 M; NaOH: 0.25 M; a: 0 g dry ice; b: 0.44 g dry ice).

TABLE 11.1
Yields of Products and Remaining Glycerine Using Dry Ice, NaHCO$_3$, and Na$_2$CO$_3$ as CO$_2$ Sources in HTW

	a	b	c	d
Lactate yield	80.1%	82.6%	81.2%	59.6%
Formate yield	0.05%	79.8%	78.1%	55.8%
Hydrogen yield	84.5%	4.0%	3.8%	3.1%
Pyruvate yield	0.64%	2.5%	2.3%	1.7%
Remaining glycerine	16.0%	14.1%	15.2%	36.7%

Note: **a**: 0.4 g NaOH; **b**: 0.4 g NaOH and 0.44 g dry ice; **c**: 0.84 g NaHCO$_3$; **d**: 1.06 g Na$_2$CO$_3$ (temperature: 300°C; reaction time: 60 min; glycerine: 0.33 M).

lactate yield. These results show that CO_2 can be reduced to formate by glycerine under hydrothermal alkaline reaction conditions.

11.3.2 The Effects of $NaHCO_3$ Quantity and NaOH Concentration

In the next series of experiments, $NaHCO_3$ was used as a CO_2 source for several reasons. First, the use of $NaHCO_3$ simplifies the experiments and allows for the accurate quantification of CO_2 [30]. In addition, when CO_2 is under alkaline conditions, it is readily converted to $NaHCO_3$, so it is necessary to study the reduction process of $NaHCO_3$ to enable the recycling of CO_2 in future industrial applications. The results in Table 11.1 show that when $NaHCO_3$ is used as a CO_2 source, the product yields are comparable to or slightly lower than the yields obtained when dry ice is used as a CO_2 source. Next, a series of experiments were performed to optimize reaction conditions by varying $NaHCO_3$ quantity, NaOH concentration, reaction temperature, and time. Because it has been reported that the addition of a base improves the reaction [41], and the conversion of glycerine into lactate is a base-catalyzed reaction [37–39], the effect of NaOH concentration on the reduction process was determined.

The effects of $NaHCO_3$ quantity and NaOH concentration on the glycerine-mediated reduction of $NaHCO_3$ under hydrothermal alkaline conditions are illustrated in Figures 11.2 and 11.3, respectively. As shown in Figure 11.2, the yields of both formate and lactate increase with higher quantities of $NaHCO_3$. The curves show a clear increase when the initial $NaHCO_3$ amount is 0.33 g or the mole ratio of $NaHCO_3$ to glycerine is approximately 3, but the curves flatten when the $NaHCO_3$ amount is 0.88 g or the ratio of $NaHCO_3$/glycerine is 8. The results indicate that the initial quantity of $NaHCO_3$ or the ratio of $NaHCO_3$/glycerine is an important factor on the glycerine-mediated reduction of $NaHCO_3$. This shows that glycerine

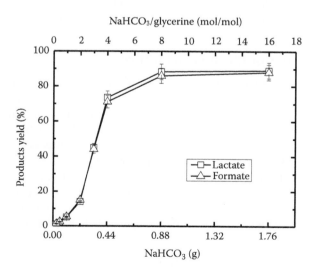

FIGURE 11.2 Effect of $NaHCO_3$ quantity on the hydrogen-transfer reduction of $NaHCO_3$ with glycerine (temperature: 300°C; reaction time: 90 min; glycerine: 0.33 M).

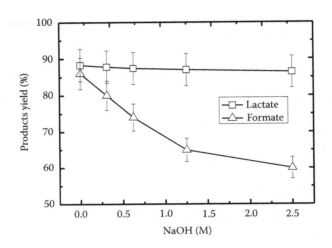

FIGURE 11.3 Effect of NaOH concentration on the hydrogen-transfer reduction of $NaHCO_3$ with glycerine (temperature: 300°C; reaction time: 90 min; glycerine: 0.33 M; $NaHCO_3$: 0.88 g).

is a stronger reductant because the optimal $NaHCO_3$/isopropanol ratio is 12 on the hydrogen-transfer reduction of $NaHCO_3$ with isopropanol [32]. Figure 11.3 shows that an increase in NaOH concentration leads to a decrease in the formate yield, but the lactate yield is almost the same with the higher NaOH concentration. This may be because more $NaHCO_3$ is converted to Na_2CO_3 at higher NaOH loadings. As shown in Table 11.1, the product yields when Na_2CO_3 is used as a CO_2 source are lower than the yields when $NaHCO_3$ is used. The lactate yield is relatively unchanged because the addition of NaOH does not substantially change the pH value of the solution.

11.3.3 The Effects of Reaction Temperature and Time

The influence of reaction temperature and time was also studied at the optimal $NaHCO_3$/glycerine ratio of 8. It can be seen in Figure 11.4 that although the yields of formate and lactate at 260°C or 280°C were very low, the yields of both compounds increased when the reaction was held at 300°C for extended reaction times. Under optimal conditions, a formate yield of approximately 90% was obtained, which is almost the same as the yield of lactate. These results show that $NaHCO_3$ can be readily reduced and converted to formate by glycerine under alkaline hydrothermal conditions; under these conditions, glycerine is converted to lactate. It should be noted that the yield of formate from $NaHCO_3$ reduction with glycerine as a reducing agent is considerably higher than the yield of formate with isopropanol in HTW [32].

11.3.4 The Catalysis Effect of Reactor Materials

In order to investigate the catalysis effect of reactor materials on the hydrogen-transfer reduction of $NaHCO_3$ with glycerine under hydrothermal alkaline conditions, we

Hydrothermal Reduction of CO_2 with Glycerine

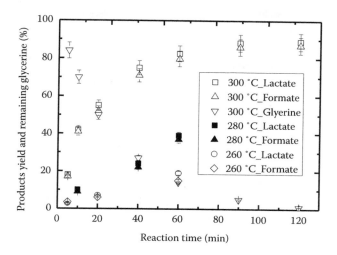

FIGURE 11.4 Effect of reaction temperature and time on the hydrogen-transfer reduction of $NaHCO_3$ by glycerine (glycerine: 0.33 M; $NaHCO_3$: 0.88 g).

performed a series of experiments with a Teflon-line batch reactor with or without the addition of a small amount of SUS316 scrap [26,39]. After 24 h at 250°C with 0.33 M glycerine and 0.88 g $NaHCO_3$, the results showed that the formate and lactate yields with or without pieces of SUS316 were almost the same. Thus, it is likely that water molecules are acting as a catalyst for the hydrogen-transfer reduction of $NaHCO_3$. To prove this, we carried out two anhydrous experiments with glycerine and dry ice at 300°C with diethylamine or NaOH as a base, and we found that neither formate nor lactate was produced. However, both formate and lactate were detected by HPLC analysis in a hydrous experiment with glycerine and dry ice at 300°C when diethylamine was used as a base. These results indicate that H_2O other than NaOH may catalyze the reaction of the hydrogen-transfer reduction of $NaHCO_3$ with glycerine.

11.3.5 THE REACTION PATHWAY ON THE HYDROGEN-TRANSFER REDUCTION OF $NaHCO_3$ WITH GLYCERINE

Finally, the reaction pathway on the hydrogen-transfer reduction of $NaHCO_3$ with glycerine was studied. As shown in Figure 11.1, a small amount of pyruvaldehyde was detected. Pyruvaldehyde is likely an intermediate product, which undergoes a benzilic rearrangement to ultimately form lactic acid. This is a known conversion in sugar chemistry and has also been observed in hydrothermal reactions at 300°C [25,42,43]. Although a detailed mechanism for the hydrogen-transfer reduction of $NaHCO_3$ with glycerine cannot yet be deduced, an outline of a potential pathway is provided in Scheme 11.1; this pathway is based on the proposed mechanism for the hydrogen-transfer reduction of $NaHCO_3$ with isopropanol [32] and takes into account the same yields of formate and lactate, the catalytic role of water molecules, and the presence of pyruvaldehyde.

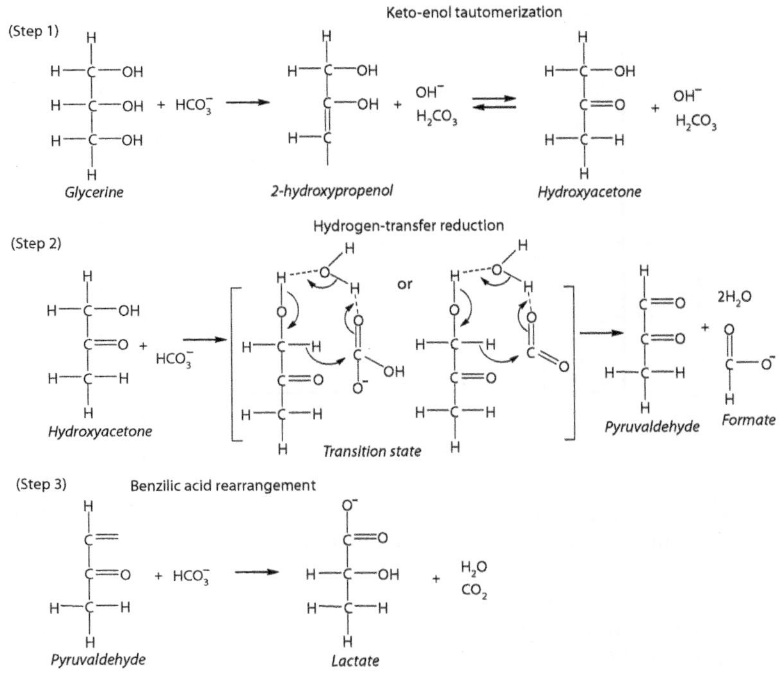

SCHEME 11.1 The proposed pathway of the hydrogen-transfer reduction of NaHCO$_3$ with glycerine.

In the first step, glycerine is dehydrated to produce 2-hydroxypropenol in E2 mechanism via a base attacking at the hydrogen of C-2 and then OH⁻ elimination of C-1. Subsequently, hydroxyacetone is formed by keto-enol tautomerization of the produced 2-hydroxypropenol. In the second step, two hydrogen bonds may be formed among three molecules (hydroxyacetone, H$_2$O, and HCO$_3^-$ or CO$_2$ from NaHCO$_3$), which makes the carbonyl-carbon on HCO$_3^-$ or CO$_2$ and the hydride ion on the hydroxyacetone even more positive. Next, the hydride ion attacks the carbonyl-carbon, and a cyclic transition state may be formed. Finally, pyruvaldehyde and formate are formed, and a water molecule is regenerated after an intramolecular hydride shift. In the third step, pyruvaldehyde undergoes a benzilic acid rearrangement to form the lactate salt. In the proposed pathway, water molecules make a hydrogen-bond ring network with the substrate molecules, and the eight-membered ring transition state greatly lowers the energy for bond cleavage and formation. Similar water-catalyzed mechanisms have been proposed in which hydrogen bonding between the substrates and water molecules forms a ring transition state in the reaction [23,24,26,44].

It is interesting that the peak of pyruvic acid in Figure 11.1b increased greatly compared to that in Figure 11.1a. As shown in columns a and b of Table 11.1, pyruvate yield increased to 2.5% from 0.64% by adding 0.44 g of CO$_2$. This finding suggests that in addition to the consecutive oxidation of pyruvaldehyde to form it [32], pyruvic acid may be mainly obtained from the hydrogen-transfer reduction of lactic

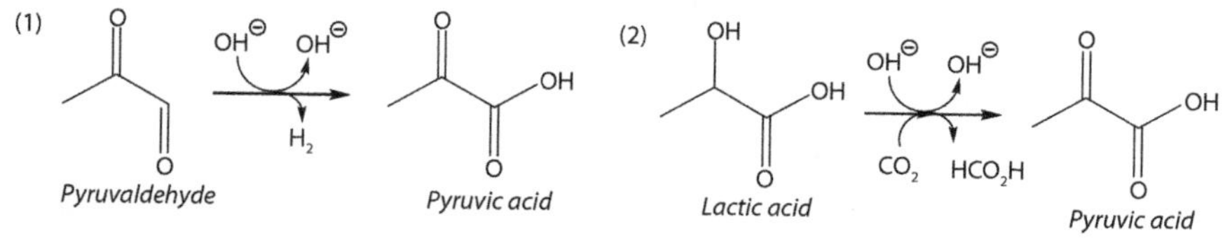

SCHEME 11.2 The proposed pathway for information of pyruvic acid as a side reaction.

acid and CO_2 because 2-hydroxyl group in lactic acid seems to act as that in isopropanol [32]. As a side reaction, this reaction pathway for information of pyruvic acid is shown in Scheme 11.2.

11.4 FROM NaHCO₃ INTO FORMATE AND FROM ISOPROPANOL INTO ACETONE: HYDROGEN-TRANSFER REDUCTION OF NaHCO₃ WITH ISOPROPANOL IN HTW

In our previous work, we have used glycerol as a high-energy starting material to produce formic acid from CO_2 under hydrothermal conditions, in which glycerol was still converted into lactic acid [45]. Glycerol is a potentially important biorefinery feedstock, available as a by-product in the production of biodiesel by transesterification of vegetable oils or animal fats [35,36]. In order to learn about the function of the hydroxyl group (R-OH) and prevent the production of carboxyl acid (R-COOH) from R-OH, isopropanol, the simplest secondary alcohol, was selected as a high-energy substance. Moreover, catalytic dehydrogenation or catalytic oxidation of isopropanol into acetone can be chosen as an alternative synthetic route although acetone is often produced as a by-product in the production of phenol [46]. Herein, we indicated that $NaHCO_3$ (CO_2 source) could be converted into formate salt using isopropanol as a reducing agent in HTW and acetone was still obtained from isopropanol, in which the effect of reaction conditions was also investigated in detail.

11.4.1 Reaction Products Analysis from Isopropanol and NaHCO₃

First, reaction products from isopropanol and $NaHCO_3$ were identified by HPLC analysis. Figure 11.5 shows the HPLC chromatogram for liquid samples after reaction with 0.25 M isopropanol and 1.0 g of $NaHCO_3$ at a temperature of 300°C and a reaction time of 10 min. It is quite evident that formate was formed after the hydrothermal reduction of isopropanol and CO_2, because a big peak of formic acid is detected.

11.4.2 The Effect of the Reaction Temperature

The effect of the reaction temperature on the hydrogen-transfer reduction of $NaHCO_3$ with isopropanol was studied by varying the temperature between 260°C and 320°C. As shown in Figure 11.6, the formate concentration increases greatly with the reaction temperature from 260°C to 300°C. The reason for the phenomenon may be that H_2 is produced more and faster at higher temperature [37], which provides a

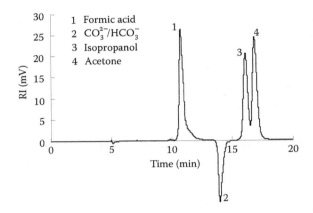

FIGURE 11.5 HPLC chromatogram for samples after hydrothermal reaction of isopropanol and NaHCO$_3$ (temperature: 300°C; reaction time: 60 min; isopropanol: 0.25 M; NaHCO$_3$: 1.0 g; RI detector).

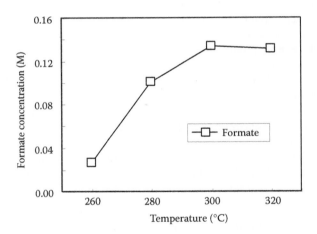

FIGURE 11.6 The effect of reaction temperature on formate concentration (reaction time: 150 min; isopropanol: 0.25 M; NaHCO$_3$: 1.0 g).

stronger reducing environment and promotes the production of formate. The formate concentration does not increase further with temperature when the temperature exceeds 300°C, which could be possibly caused by the decomposition of the formate at higher temperatures. Therefore, we preferred setting the hydrothermal reaction at a mild temperature of 300°C in the following studies.

11.4.3 The Effect of the Quantity of NaHCO$_3$

The influence of the quantity of NaHCO$_3$ on the hydrogen-transfer reduction of NaHCO$_3$ with isopropanol is shown in Figure 11.7. Formate concentration drastically

Hydrothermal Reduction of CO_2 with Glycerine

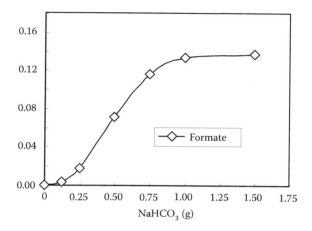

FIGURE 11.7 The effect of $NaHCO_3$ on formate concentration (temperature: 300°C; reaction time: 150 min; isopropanol: 0.25 M).

TABLE 11.2
Formate Concentration Using Dry Ice, $NaHCO_3$, and Na_2CO_3 as CO_2 Sources in HTW

	Dry Ice[a]	$NaHCO_3$[b]	Na_2CO_3[c]
Formate concentration	0.15	0.13	0.09

[a] 0.4 g of NaOH and 0.44 g of dry ice.
[b] 0.84 g of $NaHCO_3$.
[c] 1.06 g of Na_2CO_3 (temperature: 300°C; reaction time: 150 min; isopropanol: 0.25 M).

increased with increasing $NaHCO_3$ up to 1.00 g, and slowly increased when $NaHCO_3$ was further enhanced to 1.50 g. This result shows that $NaHCO_3$ has a significant effect on formate production. Because $NaHCO_3$ as a CO_2 source could be converted into CO_2 in HTW, $NaHCO_3$ were compared with CO_2 by adding dry ice in the presence of NaOH. The results in Table 11.2 show that the formate concentration produced using dry ice is comparable to or slightly higher than that using $NaHCO_3$. It is possible that the solubility of CO_2 in liquid water initially decreases as the temperature is increased above ambient, but a minimum is soon reached around 150°C, and then the gas solubility increases [22].

11.4.4 The Effect of the NaOH Concentration

Because it has been recognized that the base is beneficial to improving enthalpy of the reduction of CO_2 [41], the effect of NaOH concentration was investigated on the hydrogen-transfer reduction of $NaHCO_3$ with isopropanol. As shown in Figure 11.8,

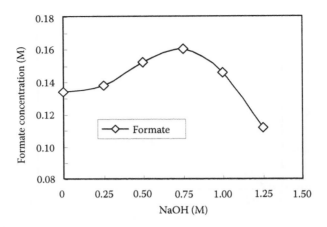

FIGURE 11.8 The effect of NaOH concentration on formate concentration (temperature: 300°C; reaction time: 150 min; isopropanol: 0.25 M; NaHCO$_3$: 1.0 g).

formate concentration passed a maximum at NaOH concentration of 0.75 M. An excess NaOH results in a decrease of formate concentration from the maximum of 0.16 M, which may be caused by the fact that more NaHCO$_3$ was converted into Na$_2$CO$_3$ with the addition of excess NaOH. As shown in Table 11.2, the formate concentration using Na$_2$CO$_3$ as a CO$_2$ source was less than that using NaHCO$_3$.

11.4.5 The Effect of the Reaction Time

The effect of reaction time on the hydrogen-transfer reduction of NaHCO$_3$ with isopropanol is shown in Figure 11.9. It can be seen in Figure 11.9 that the remaining isopropanol gradually decreased with the increase of reaction time. The formate

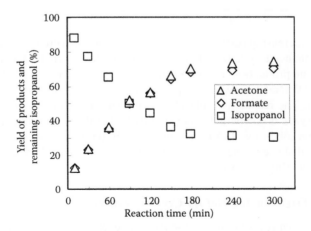

FIGURE 11.9 The effect of reaction time on hydrogen-transfer reduction of NaHCO$_3$ with isopropanol (temperature: 300°C; isopropanol: 0.25 M; NaHCO$_3$: 1.0 g; NaOH: 0.75 M).

yield increases correspondingly with the increasing of reaction time, and a high level of formate yield (70%) is achieved. The yield of products is defined by the mole of products divided by that of the initial reducing agent, isopropanol. It could be noted that formate yield from CO_2 using isopropanol as a reducing agent was considerably higher than formate yield using metal-reducing agents in HTW [27–30]. This result suggested that organic hydrogen source as reducing agents seems to have higher efficiency for reducing CO_2 than that with metal-reducing agents under hydrothermal conditions.

11.4.6 THE CATALYSIS EFFECT OF REACTOR MATERIALS

Then, the catalysis effect of reactor materials was investigated using a Teflon-lined batch reactor with and without the addition of a slight amount of SUS 316 scraps. Results showed that after 24 h at 250°C with 0.25 M isopropanol and 1.0 g $NaHCO_3$, the yields of formate obtained with and without SUS 316 were almost the same. Then, water molecules may act as a catalyst. To prove this, we carried out two additional anhydrous experiments with isopropanol and dry ice at 300°C using dialkylamines or using NaOH as a base, and the results found that formic acid or formate was not produced. However, formate was detected by the HPLC analysis from a hydrous experiment with isopropanol and dry ice at 300°C using dialkylamines as a base. These results indicated the catalytic role of water molecules on the hydrogen-transfer reduction of $NaHCO_3$ with isopropanol. It has been reported that addition of small amounts of water in organic solvent could accelerate reduction of CO_2 [41,47]. Moreover, Nguyen and Ha reported that water and CO_2 interacting in such a way have been calculated by ab initio methods to be more stable than the two species apart [48].

11.4.7 THE REACTION PATHWAY ON THE HYDROGEN-TRANSFER REDUCTION OF $NaHCO_3$ WITH ISOPROPANOL

On the basis of the traditional Meerwein–Ponndorf–Verley (MPV) hydrogen-transfer reduction mechanism involving a cyclic transition state and the catalytic role of water molecules, a possible mechanism was proposed, although a detailed mechanism for hydrogen-transfer reduction of $NaHCO_3$ with isopropanol cannot yet be deduced. As shown in Scheme 11.1, first, two hydrogen bondings may be formed among three molecules (isopropanol, H_2O, and HCO_3^- or CO_2 from decomposition of $NaHCO_3$), making carbonyl-carbon of HCO_3^- or CO_2 and hydride of isopropanol even more positive. Then, the hydride of isopropanol attacks the carbonyl-carbon of HCO_3^- or CO_2 and a cyclic transition state may be formed. Finally, after the intramolecular hydride shift, formate and acetone were produced, and water molecules were recovered.

According to the possible mechanism in Scheme 11.3, the yield of formate should be equal to that of acetone. To test this, acetone in liquid compositions was analyzed. As shown in Figure 11.9, the formate and acetone were generated equally at the initial stage, whereas acetone was more than formate at the final stage, suggesting that the prolonged reaction time resulted in the more decomposition of formate than acetone. Additionally, it has been long considered that hydrogen-transfer reactions

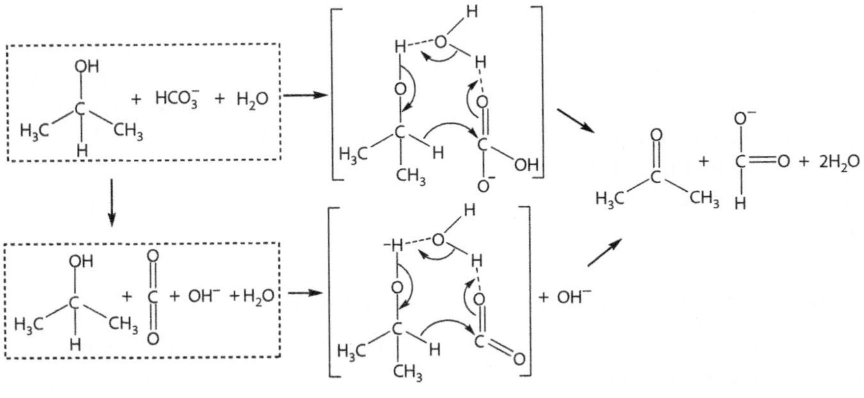

SCHEME 11.3 The possible mechanism for hydrogen-transfer reduction of NaHCO$_3$ with isopropanol.

are reversible, such as MPV reduction and Oppenauer oxidation. Our previous work indicated that acetone and formic acid produced isopropanol and CO$_2$ by a hydrogen-transfer reduction reaction at 300°C, and its mechanism was also the reverse as shown in Scheme 11.1 [26,44]. Similarly, Farlow and Adkins in 1935 reported the first direct synthesis of formic acid from CO$_2$ and H$_2$ using Raney nickel catalyst (Equation 11.1: CO$_2$ + H$_2$ → HCOOH) [10], and Inoue et al. in 1976 discovered that the formic acid decomposed to CO$_2$ and H$_2$ once the pressures of these gases are reduced because catalysts for Reaction 11.1 are also catalysts for the reverse reaction [41]. These results strongly support the possible mechanism shown in Scheme 11.1 for the hydrogen-transfer reduction of NaHCO$_3$ with isopropanol.

11.5 THE MECHANISM FOR PRODUCTION OF ABIOGENIC FORMATE FROM CO$_2$ AND LACTATE FROM GLYCERINE: UNCATALYZED TRANSFER HYDROGENATION OF CO$_2$ WITH GLYCERINE UNDER ALKALINE HYDROTHERMAL CONDITIONS

Biomass is an abundant source of alcohols in the form of carbohydrates and polyols such as cellulose, starch, and glycerine [49,50]. Moreover, glycerine has been a potentially important biorefinery feedstock as a by-product of biodiesel production. Our recent research showed that, in HTW, CO$_2$ is reduced to formate by the alcohol-mediated reduction using isopropanol or glycerine as alcohol model compounds [32,51]. Like hydrogen-transfer reduction of CO$_2$ with H$_2$, these reactions are endergonic (ΔGRT > 0), but a hydrogen transfer reduction of CO$_2$ with glycerine requires much less energy (ΔGRT < 0) (Equations 11.1 and 11.4), and the addition of a base improves the enthalpy of the reaction, while dissolution of the gases improves the entropy (Equations 11.2, 11.3, 11.5, and 11.6). We found that CO$_2$ could be effectively converted into formate using glycerine as a reducing agent, and the molar yield of formate was almost equal to that of lactate from glycerine [51]. However, little

Hydrothermal Reduction of CO₂ with Glycerine

is known about the mechanism for production of abiogenic formate from CO_2 and lactate from glycerine during such processes. More recently, we have also proven that glycerine was first transformed to acetol via a dehydration reaction and a keto-enol tautomerization reaction during the production of hydrogen and lactic acid from glycerine [52]. Herein, we investigated the formation of abiogenic formate from CO_2, D_2O solvent effect, reactor materials effect, and H_2O molecules catalysis for uncatalyzed transfer hydrogenation of CO_2 with glycerine under alkaline hydrothermal conditions, and then have proposed and have proven the potential reaction mechanism. The present work should help facilitate studies on industrial application of CO_2 reduction with abundant alcohol compounds as reducing materials rather than hydrogen and the development of renewable high-valued chemicals from alternative biomass derivatives and the primary greenhouse gas to fossil fuel.

$$CO_{2(g)} + H_{2(g)} \xrightarrow{32.9 \text{ kJ/mol}} HCOOH_{(l)} \tag{11.1}$$

$$CO_{2(g)} + H_{2(g)} + NH_{3(aq)} \xrightarrow{-9.5 \text{ kJ/mol}} HCOO^-_{(aq)} + NH^+_{4(aq)} \tag{11.2}$$

$$CO_{2(aq)} + H_{2(aq)} + NH_{3(aq)} \xrightarrow{-35.4 \text{ kJ/mol}} HCOO^-_{(aq)} + NH^+_{4(aq)} \tag{11.3}$$

$$C_3H_5(OH)_3 + CO_{2(g)} \xrightarrow{-13 \text{ kJ/mol}} CH_3CH(OH)COOH_{(aq)} + HCOOH_{(l)} \tag{11.4}$$

$$C_3H_5(OH)_3 + CO_{2(g)} + NH_{3(aq)} \xrightarrow{-55.3 \text{ kJ/mol}} CH_3CH(OH)COOH_{(aq)} + HCOO^-_{(aq)} + NH^+_{4(aq)} \tag{11.5}$$

$$C_3H_5(OH)_3 + CO_{2(aq)} + NH_{3(aq)} \xrightarrow{-81.2 \text{ kJ/mol}} CH_3CH(OH)COOH_{(aq)} + HCOO^-_{(aq)} + NH^+_{4(aq)} \tag{11.6}$$

11.5.1 Formation of Abiogenic Formate from CO_2

Figure 11.10 shows the ¹H- and ²H-NMR spectra of the solutions after the hydrothermal reactions of glycerine with NaOH and CO_2 in H_2O (or NaOD and CO_2 in D_2O) at 300°C. From a comparison of the time-dependent spectra of Figure 11.10a, b, and c, it can be seen that the gradual consumption of the reactant glycerine was accompanied by the production of high-valued lactate and formate. The transformation of glycerine into lactate was almost completed in 90 min under alkaline hydrothermal conditions, and the increase in reaction time from 30 min to 90 min resulted in a monotonous increase in the production of formate. Although a set of comparative

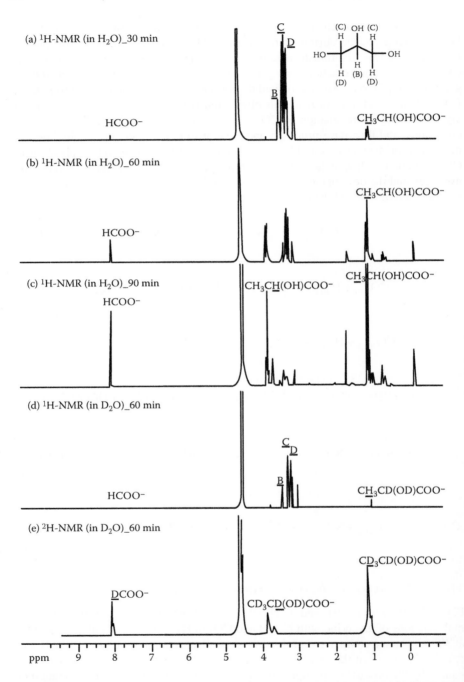

FIGURE 11.10 ^{1}H-NMR spectra for the solution after the hydrothermal reaction of 0.33 M glycerine at 300°C with 0.40 g of NaOH and 0.44 g of CO_2 in H_2O for (a) 30 min, (b) 60 min, (c) 90 min, and (d) ^{1}H-NMR and (e) ^{2}H-NMR spectra with 0.41 g of NaOD and 0.44 g of CO_2 in D_2O for 60 min.

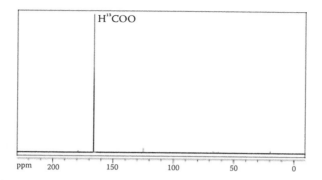

FIGURE 11.11 ^{13}C-NMR spectra for the solution after the hydrothermal reaction of 0.33 M glycerine at 300°C with 0.89 g of NaH^{13}CO$_3$ in H$_2$O for 90 min.

experiments with or without CO$_2$ source suggested that the formate was produced from CO$_2$ in Figure 11.1, it might be from organic acids because it has been reported that formic acid can be produced by decomposition of lactic acid [53,54]. Thus, in order to further acknowledge the production of abiogenic formate from CO$_2$, an experiment was carried out by using 0.89 g NaH^{13}CO$_3$ as a CO$_2$ source with 0.33 M glycerine in H$_2$O at 300°C for 90 min. The collected liquid sample was adjusted with sulfuric acid until the pH values of the solution reached 2–3 and then was detected by ^{13}C-NMR analysis. As shown in Figure 11.11, the produced formate (H^{13}COO$^-$) was observed at 165 parts per million. These results suggested that CO$_2$ was indeed converted into abiogenic formate, and at the same time glycerine was transformed into lactate during such processes.

11.5.2 Effect of D$_2$O Solvent

To investigate the solvent isotope effect during the production of abiogenic formate from CO$_2$ and that of lactate from glycerine under alkaline hydrothermal conditions, we carried out a deuterium transformation study using 0.41 g of NaOD and 0.44 g of CO$_2$ in D$_2$O instead of 0.40 g of NaOH and 0.44 g of CO$_2$ in H$_2$O [52], and the results are shown in the ^{1}H- and ^{2}H-NMRs in Figure 11.10d and e, respectively. By comparing Figure 11.10a and d to Figure 11.10e, it can be observed that the Hs on the β-C of lactate has been almost transformed into D when in D$_2$O, and the remaining glycerine did not participate in the H/D exchange reaction. However, it has been reported that simple alcohols do not participate in H/D exchange reactions, whereas the α-C of carboxyl groups undergo rapid and nearly complete exchange [55,56]. From these results, we can speculate that there is an intermediate product, such as R$_1$-CO-R$_2$, formed during the production of lactate from glycerine because of the H/D exchange on the β-C of lactate.

To further understand the deuterium behavior in the formation of lactate from glycerine in D$_2$O under alkaline hydrothermal conditions, the liquid samples obtained from the reaction of 0.41 g of NaOD and 0.44 g of CO$_2$ in D$_2$O at a temperature of 300°C were also analyzed by LC-MS. As shown in Figure 11.12a1 and b1, the m/z of

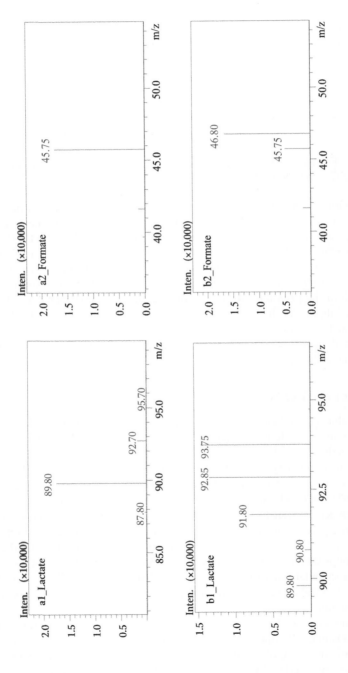

FIGURE 11.12 MS spectra of the lactate and formate solutions after the hydrothermal reactions of 0.33 M glycerine at 300°C with (a) 0.40 g of NaOH and 0.44 g of CO_2 in H_2O and (b) 0.41 g of NaOD and 0.44 g of CO_2 in D_2O.

lactate after reacting in H_2O was 89.80, but that after reacting in D_2O increased to 91.80, 92.85, and 93.75. This result from the LC-MS analysis is consistent with that of the previous NMR analysis, where a large number of Hs on the β-Cs of lactate were indeed exchanged by D in D_2O. These results suggest that the hydroxyl (–OH) group on the 2-C of glycerine converted to a carbonyl (C=O) group and then was reverted into an –OH group in the α-C of lactate.

11.5.3 Effect of Reactor Materials

It is well known that the SUS 316 material contains some metals, such as Fe, Ni, Mo, and Cr. Recently, it was reported that formic acid can be produced from CO_2 via the oxidation of a zero-valent metal under hydrothermal conditions [27–31,33]. Hence, metals in the SUS 316 reactor used in this study may play a catalytic role during the transfer hydrogenation of CO_2 into formate with glycerine under alkaline hydrothermal conditions. In order to investigate the catalysis effect of reactor materials on the transfer hydrogenation of CO_2 with glycerine under hydrothermal alkaline conditions, we performed a series of experiments with a Teflon-line batch reactor with or without the addition of a small amount of SUS316 scrap. After 24 h at 200°C with 0.33 M glycerine with 1.40 g of NaOH and 1.54 g of CO_2, the results showed that the formate and lactate yields with or without pieces of SUS316 were almost the same, as shown in Table 11.3. This result suggests that the catalytic effect of the reactor material is not obvious.

11.5.4 Catalysis of H_2O Molecules

It is likely that water molecules are acting as a catalyst for the transfer hydrogenation of CO_2 with glycerine under alkaline hydrothermal conditions because of the absence of significant catalytic effects of the reactor material. To prove this, we carried out two anhydrous experiments with glycerine and dry ice at 300°C using diethylamine or NaOH as a base, and we found that neither formate nor lactate was produced. However, both formate and lactate were detected by HPLC analysis in a hydrous experiment with glycerine and dry ice at 300°C when diethylamine was used as a base. These results indicate that H_2O other than NaOH may catalyze the

TABLE 11.3
Yields of Products with or without Pieces of SUS316 after 24 h at 200°C with 0.33 M Glycerine with 0.40 g of NaOH and 0.44 g of CO_2 in a Teflon-Line Batch Reactor

	Lactate (%)	Formate (%)
Without pieces of SUS316	78.8	77.0
With pieces of SUS316	79.5	78.2

reaction of the transfer hydrogenation of CO_2 with glycerine under alkaline hydrothermal conditions. It has also been reported that addition of small amounts of water in organic solvent could accelerate the reduction of CO_2 [41,47]. Moreover, Nguyen and Ha reported that water and CO_2 interacting in such a way have been calculated by ab initio methods to be more stable than the two species apart [48].

11.5.5 Proposition of Reaction Mechanism

An outline of a potential mechanism on the transfer hydrogenation of CO_2 with glycerine under alkaline hydrothermal conditions is provided, on the basis of the above isotope effect of solvent, the catalytic role of water molecules, the almost same yields of lactate and formate, and the detected pyruvaldehyde in Figure 11.1. As shown in Scheme 11.4, in the first step, glycerine is dehydrated to produce 2-hydroxypropenol in E2 mechanism via a base attacking the hydrogen of C-2 and then OH^- elimination of C-1. Subsequently, acetol is formed by keto-enol tautomerization of the produced 2-hydroxypropenol. In the second step, two hydrogen bonds may be formed among three molecules (acetol, H_2O, and CO_2), which makes the carbonyl-carbon on CO_2 and the hydride ion on the acetol even more positive. Next, the hydride ion attacks the carbonyl-carbon, and a cyclic transition state may be formed. Finally, pyruvaldehyde and formate are formed, and a water

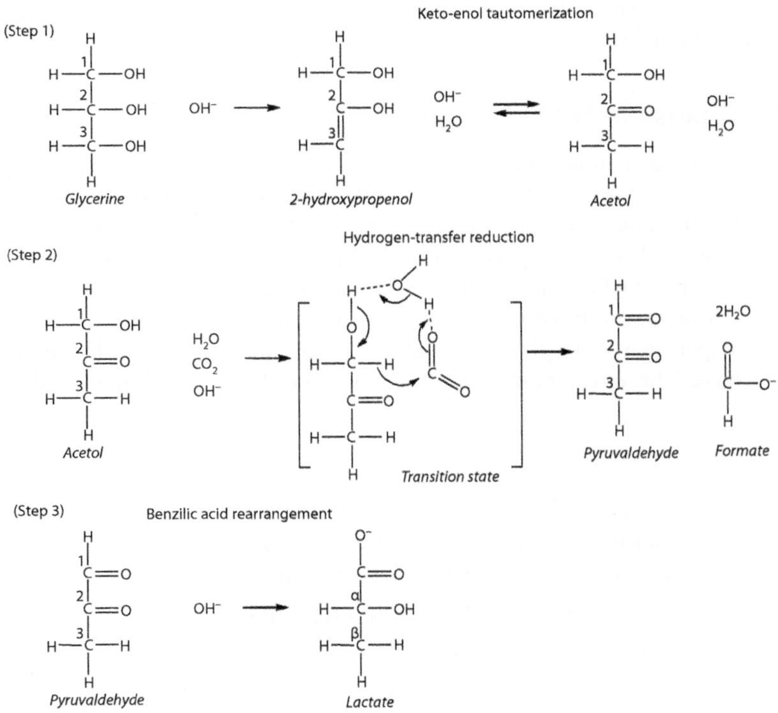

SCHEME 11.4 Potential mechanism for transfer hydrogenation of CO_2 with glycerine from glycerine under hydrothermal conditions.

molecule is regenerated after an intramolecular hydride shift. In the third step, pyruvaldehyde undergoes a benzilic acid rearrangement to form the lactate salt. In the proposed pathway, water molecules make a hydrogen-bond ring network with the substrate molecules, and the eight-membered ring transition state greatly lowers the energy for bond cleavage and formation. Similar water-catalyzed mechanisms have been proposed in which hydrogen bonding between the substrates and water molecules forms a ring transition state in the reaction [23,24,55]. Furthermore, it was reported that uncatalyzed MPV reduction could be achieved under hydrothermal conditions by using water molecules as catalysts [57–59].

Additionally, it has long been considered that hydrogen-transfer reactions are reversible, such as MPV reduction and Oppenauer oxidation. Our previous work indicated that reaction of acetone and formic acid and reaction of isopropanol and CO_2 are a pair of reversible reactions [26,32]. Similarly, Farlow and Adkins in 1935 reported the first direct synthesis of formic acid from CO_2 and H_2 using Raney nickel catalyst (Equation 11.7: $CO_2 + H_2 \rightarrow HCOOH$) [10], and Inoue et al. in 1976 discovered that the formic acid decomposed to CO_2 and H_2 once the pressures of these gases are reduced because catalysts for Reaction 11.7 are also catalysts for the reverse reaction [41]. On the transfer hydrogenation of CO_2 with glycerine under alkaline hydrothermal conditions, however, after hydrogen-transfer reactions, the formed pyruvaldehyde was further transformed into more stable lactate via a benzilic acid rearrangement, which thereby slows the inverse reaction and then achieves a higher lactate yield.

11.5.6 Testing of Reaction Mechanism

From the postulated mechanism shown in Scheme 11.4, we can presume that H_2 decreases substantially because the hydride ion attacks the CO_2 better than H_2O by adding CO_2. As shown in columns a through c of Table 11.1, H_2 yields decreased to 4.0% and 3.8% from 84.5% by adding CO_2 or $NaHCO_3$. Moreover, from the postulated mechanism shown in Scheme 11.4, we can presume that H/D exchange was achieved in C-3 position because of the ketone carbonyl group of acetol similar to the C-1 position of acetol. In an independent run, we confirmed that $HCOO^-$ cannot be transformed into $DCOO^-$ in D_2O under solely alkaline hydrothermal conditions. From these above, we can predict that in D_2O reactions, the produced formate included $DCOO^-$ and $HCOO^-$, and furthermore $DCOO^-$ was more than $HCOO^-$, because of rapid and nearly complete H/D exchange under hydrothermal conditions [55,56]. Thus, in order to detect D in the produced formate, the liquid sample obtained from the reaction of glycerine with NaOD and CO_2 in D_2O (or NaOH and CO_2 in H_2O) at a temperature of 300°C was analyzed by ^{2}H-NMR and LC-MS. By comparing Figure 11.10b and d to Figure 11.10e, it can be observed that the Hs on formate have been almost transformed into Ds when in D_2O solvent. This result from the H-NMR analysis is consistent with that of the next LC-MS analysis. Uniformly, it can be seen in Figure 11.12a2 and b2 by LC-MS analyses that the *m/z* of formate after reacting in H_2O was 45.75, but that after reacting in D_2O, it was 45.75 and 46.80. The presence of large amounts of D in the produced formate straightforwardly indicates that acetol was formed in the first place as the most probable intermediate.

TABLE 11.4
Comparison of the Lactate Yields from Pyruvaldehyde and Acetol

	Pyruvaldehyde[a]	Acetol[b]
Lactate (%)	96.7	59.6
Formate (%)	–	42.3
Remaining (%)	Trace	Trace

[a] Pyruvaldehyde 0.33 M; NaOH 1.25 M; CO_2 0.44 g; temperature 300°C; time 1.5 min.
[b] Acetol 0.33 M; NaOH 1.25 M; CO_2 0.44 g; temperature 300°C; time 1.5 min.

According to the proposed mechanism shown in Scheme 11.4, pyruvaldehyde and acetol could be regarded as the two key intermediates on transfer hydrogenation of CO_2 with glycerine under hydrothermal alkaline conditions. Therefore, we investigated the possibilities for producing lactate from pyruvaldehyde and acetol under hydrothermal alkaline conditions. As shown in Table 11.4, the yield for lactate from 0.33 M pyruvaldehyde was 96.7% at 300°C with 1.25 M NaOH and 0.44 g of CO_2 for a reaction time of 1.5 min. Additionally, when an experiment with 0.33 M acetol, 1.25 M NaOH, and 0.44 g of CO_2 was performed under hydrothermal conditions at 300°C after 1.5 min, the yields of lactate and formate were only 59.6% and 42.3%, respectively, as shown in Table 11.4. In the reaction of acetol and CO_2, the yields of lactate and formate were relatively low because of the impact of the aldol condensation of acetol. Table 11.1 corroborates this presumption by showing lactate to be the major product obtained from pyruvaldehyde and acetol under these conditions.

11.5.7 Verification of Alkaline Role

Although the dehydration and keto-enol tautomerization reactions occur regularly in acid-catalyzed conditions, these reactions have also been reported in base-catalyzed conditions [60–62]. The benzilic acid rearrangement occurs only in a basic solution. Furthermore, it has been recognized that addition of a base improves enthalpy of the reduction reaction of CO_2 and dissolution of the gases improves the entropy (Equations 11.2, 11.3, 11.5, and 11.6) [41,47]. Thus, the transfer hydrogenation of CO_2 with glycerine under alkaline hydrothermal conditions must be critically dependent on the concentration of the base. To confirm this alkaline role, experiments were performed with glycerine and 0.44 g of CO_2 at 300°C, varying the NaOH concentration from 0 to 1.25 M. As shown in Figure 11.13, both lactate and formate were not formed when the NaOH concentration was 0 M. However, the lactate and formate yields did increase sharply with an increase in the NaOH concentration. These results strongly support the proposed mechanism for transfer hydrogenation of CO_2 with glycerine under alkaline hydrothermal conditions shown in Scheme 11.4.

Hydrothermal Reduction of CO_2 with Glycerine

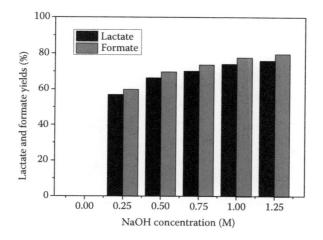

FIGURE 11.13 Lactate and formate yields after the hydrothermal reactions of 0.33 M glycerine and 0.44 g of dry ice with 0–2.5 M NaOH at 300°C for 60 min.

11.5.8 Alkaline Hydrothermal Conversion Routes of Lactate and Intermediates

As shown in the proposed mechanism in Scheme 11.4, 1 mol glycerine and 1 mol CO_2 would generate 1 mol lactate and 1 mol formate through the following stoichiometric reaction (Equation 11.8):

$$C_3H_8O_3 + CO_2 + 2OH^- \rightarrow C_3H_5O_3^- + HCOO^- + 2H_2O \qquad (11.8)$$

HPLC chromatograms of the reaction mixture after a reaction is carried out at 300°C, with glycerine and 0.25 M NaOH with or without dry ice (CO_2 source) for 60 min, are shown in Figure 11.1. As it can be seen in Figure 11.1, pyruvate, pyruvaldehyde, acrylate, formate, and acetate were detected in addition to lactate. According to the proposed mechanism in Scheme 11.4, acetol and pyruvaldehyde are intermediates on the hydrogen-transfer reduction of CO_2 with glycerine, but acetol was not detected in Figures 11.1 and 11.10. Based on all these findings, several alkaline hydrothermal conversion routes of lactate and reaction intermediates can be envisioned, which are given in Scheme 11.5.

These routes are not intended to be the only ones existing, but they are several of the more probable ones. In route 1, under hydrothermal alkaline conditions, lactate was firstly decarbonized to produce acetaldehyde, which further oxidated into acetate by H_2O and CO_2 through ethanol and acetaldehyde as intermediates. In routes 2 and 3, formate and acetate were finally formed from the oxidative cleavage of acetol and pyruvaldehyde under alkaline conditions through methanol, formaldehyde, and acetaldehyde as intermediates. In route 4, acrylate was produced from dehydration of lactate, further decarbonized to form ethene, and then oxidized into acetate by H_2O and CO_2 through ethanol and acetaldehyde as intermediates. In route 5, the pyruvate

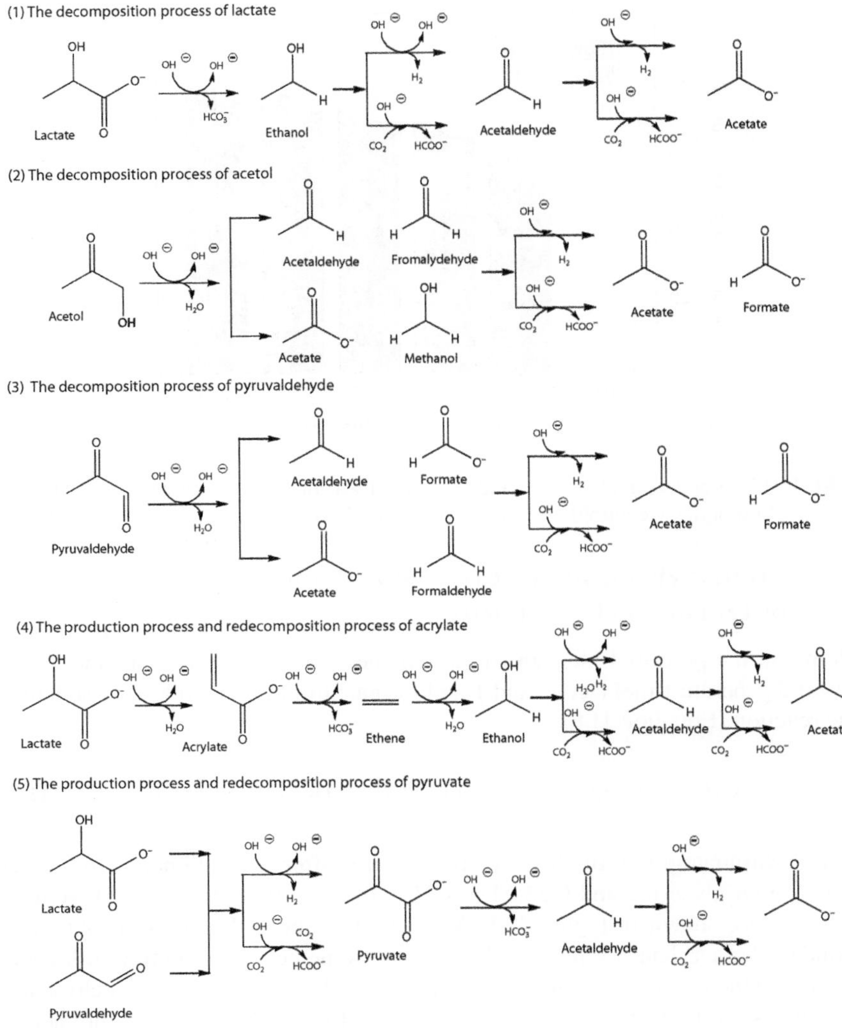

SCHEME 11.5 Alkaline hydrothermal conversion routes of lactate, aectol, pyruvaldehyde, acrylate, and pyruvate.

produced from the oxidation of pyruvaldehyde and lactate under alkaline conditions was further decarbonized into acetaldehyde and was finally oxidated into acetate by H_2O and CO_2.

From the proposed mechanism in Scheme 11.1 and routes 1 to 5 in Scheme 11.2, we could deduce that addition of CO_2 can accelerate the conversion of glycerine into lactate and slow down the decarbonization of oxidative cleavage of C3 compounds into C1 and C2 compounds. As can be seen from Table 11.1, acetate yield was greatly decreased by adding CO_2 or $NaHCO_3$. Although formate yield was rapidly increased owing to the reduction of CO_2, it suggests that, on the contrary, the amount of formate from decomposition of C3 compounds was affirmatively decreased. Interestingly, the peak area of pyruvate in Figure 11.1b increased greatly compared to that in Figure 11.1a, and that of acrylate relatively decreased. As shown in columns a and b of Table 11.1, pyruvate yield increased to 2.5% from 0.64% by adding 0.44 g of CO_2, and acrylate yield decreased to 0.5% from 2.2%. It is attributed to the fact that although pyruvate was formed by the consecutive oxidation from pyruvaldehyde in addition to that from lactate [42], pyruvate may be mainly obtained from the hydrogen-transfer reduction of lactate and CO_2 because the 2-hydroxyl group in lactate seems to act the same as that in acetol. As side reactions of routes 1 to 5 in Scheme 11.5, these results also support the reaction mechanism in Scheme 11.4.

11.6 CONCLUSIONS

1. We report a hydrothermal method for the hydrogen-transfer reduction of CO_2 or $NaHCO_3$ that offers some significant advantages over existing methods. The reported method uses glycerine as a reducing agent and permits the conversion of $NaHCO_3$ into formate in approximately a 90% yield with concomitant production of a lactate salt. Another interesting aspect of this system is the simultaneous conversion of waste glycerine into lactic acid, which can be an intermediate for biodegradable polymers.
2. The hydrothermal method for hydrogen-transfer reduction of $NaHCO_3$ presented here offers some significant advantages, since it uses isopropanol as a reducing agent and permits reduction of $NaHCO_3$ into formate. Based on the initial reducing agent isopropanol, a higher yield of formate (approximately 70% than that for metal-reducing agents) was obtained. Another curious aspect in HTW is that water molecules may act as a catalyst on the hydrogen-transfer reduction of $NaHCO_3$ with isopropanol rather than metal catalysts in heterogeneous and homogeneous catalytic hydrogenations of CO_2.
3. We have reported the formation of abiogenic formate from CO_2, D_2O solvent effect, reactor materials effect, and H_2O molecules catalysis for uncatalyzed transfer hydrogenation of CO_2 with glycerine under alkaline hydrothermal conditions. Abiogenic formate was formed by the transfer hydrogenation of CO_2 with glycerine, and glycerine was almost completely converted into lactate with the same excellent yield of formate. A discussion on the potential mechanism for the transfer hydrogenation of CO_2 with glycerine suggests that, in the first step, glycerine is converted into acetol via dehydration and keto-enol tautomerization. In the second step, two hydrogen bonds may be formed among three molecules (acetol, H_2O, and CO_2), which makes the carbonyl-carbon on CO_2 and the hydride ion on the acetol even more positive. Next, the hydride ion attacks the carbonyl-carbon, and a cyclic transition state may be formed. Finally, pyruvaldehyde and formate are formed, and a water molecule is regenerated after an intramolecular hydride shift. In the third step, pyruvaldehyde undergoes a benzilic acid rearrangement to form the lactate salt.
4. The present work should help facilitate studies on the industrial application of CO_2 reduction with abundant alcohol compounds as reducing materials

rather than hydrogen and the development of renewable high-valued chemicals from alternative biomass derivatives and the primary greenhouse gas to fossil fuel.

REFERENCES

1. Arakawa, H., Aresta, M., Armor, J. N., Barteau, M. A., Beckman, E. J., Bell, A. T., Bercaw, J. E., Creutz, C., Dinjus, E., Dixon, D. A., Domen, K., DuBois, D. L., Eckert, J., Fujita, E., Gibson, D. H., Goddard, W. A., Goodman, D. W., Keller, J., Kubas, G. J., Kung, H. H., Lyons, J. E., Manzer, L. E., Marks, T. J., Morokuma, K., Nicholas, K. M., Periana, R., Que, L., Rostrup-Nielson, J., Sachtler, W. M. H., Schmidt, L. D., Sen, A., Somorjai, G. A., Stair, P. C., Stults, B. R., and Tumas, W. 2001. Catalysis research of relevance to carbon management: Progress, challenges, and opportunities. *Chemical Reviews* 101:953–996.
2. Sakakura, T., Choi, J.-C., and Yasuda, H. 2007. Transformation of carbon dioxide. *Chemical Reviews* 107 (6):2365–2387.
3. Darensbourg, D. J. 2007. Making plastics from carbon dioxide: Salen metal complexes as catalysts for the production of polycarbonates from epoxides and CO_2. *Chemical Reviews* 107 (6):2388–2410.
4. Mikkelsen, M., Jørgensen, M., and Krebs, F. C. 2010. The teraton challenge. A review of fixation and transformation of carbon dioxide. *Energy & Environmental Science* 3 (1):43–81.
5. Ohishi, T., Nishiura, M., and Hou, Z. 2008. Carboxylation of organoboronic esters catalyzed by n-heterocyclic carbene copper (I) complexes. *Angewandte Chemie—German Edition* 120 (31):5876.
6. Eghbali, N., and Li, C.-J. 2007. Conversion of carbon dioxide and olefins into cyclic carbonates in water. *Green Chemistry* 9 (3):213–215.
7. Jessop, P. G., Joó, F., and Tai, C.-C. 2004. Recent advances in the homogeneous hydrogenation of carbon dioxide. *Coordination Chemistry Reviews* 248 (21):2425–2442.
8. Aida, T., and Inoue, S. 1996. Metalloporphyrins as initiators for living and immortal polymerizations. *Accounts of Chemical Research* 29 (1):39–48.
9. Gibson, D. H. 1996. The organometallic chemistry of carbon dioxide. *Chemical Reviews* 96 (6):2063–2096.
10. Farlow, M. W., and Adkins, H. 1935. The hydrogenation of carbon dioxide and a correction of the reported synthesis of urethans. *Journal of the American Chemical Society* 57 (11):2222–2223.
11. Munshi, P., Main, A. D., Linehan, J. C., Tai, C.-C., and Jessop, P. G. 2002. Hydrogenation of carbon dioxide catalyzed by ruthenium trimethylphosphine complexes: The accelerating effect of certain alcohols and amines. *Journal of the American Chemical Society* 124 (27):7963–7971.
12. Jessop, P. G., Hsiao, Y., Ikariya, T., and Noyori, R. 1996. Homogeneous catalysis in supercritical fluids: Hydrogenation of supercritical carbon dioxide to formic acid, alkyl formates, and formamides. *Journal of the American Chemical Society* 118 (2):344–355.
13. Leitner, W., Dinjus, E., and Gaßner, F. 1994. Activation of carbon dioxide: IV. Rhodium-catalysed hydrogenation of carbon dioxide to formic acid. *Journal of Organometallic Chemistry* 475 (1–2):257–266.
14. Ogo, S., Kabe, R., Hayashi, H., Harada, R., and Fukuzumi, S. 2006. Mechanistic investigation of CO2 hydrogenation by Ru (II) and Ir (III) aqua complexes under acidic conditions: Two catalytic systems differing in the nature of the rate determining step. *Dalton Transactions* (39):4657–4663.

15. Urakawa, A., Jutz, F., Laurenczy, G., and Baiker, A. 2007. Carbon dioxide hydrogenation catalyzed by a ruthenium dihydride: A DFT and high-pressure spectroscopic investigation. *Chemistry—A European Journal* 13 (14):3886–3899.
16. Weber, M., Wang, J.-T., Wasmus, S., and Savinell, R. F. 1996. Formic acid oxidation in a polymer electrolyte fuel cell—A real-time mass-spectrometry study. *Journal of the Electrochemical Society* 143 (7):L158–L160.
17. Akiya, N., and Savage, P. E. 1998. Role of water in formic acid decomposition. *AIChE Journal* 44 (2):405–415.
18. Rice, C., Ha, S., Masel, R. I., Waszczuk, P., Wieckowski, A., and Barnard, T. 2002. Direct formic acid fuel cells. *Journal of Power Sources* 111 (1):83–89.
19. Uhm, S., Chung, S. T., and Lee, J. 2008. Characterization of direct formic acid fuel cells by impedance studies: In comparison of direct methanol fuel cells. *Journal of Power Sources* 178 (1):34–43.
20. Thomas, R. W., Antony, B. B., Charles, A. C., and Franck, A. E. 1991. Supercritical water: A medium for chemistry. *Chemical Engineering News* 69:26–39.
21. Akiya, N., and Savage, P. E. 2002. Roles of water for chemical reactions in high-temperature water. *Chemical Reviews* 102 (8):2725–2750.
22. Hunter, S. E., and Savage, P. E. 2008. Quantifying rate enhancements for acid catalysis in CO_2-enriched high-temperature water. *AIChE Journal* 54 (2):516–528.
23. Takahashi, H., Hisaoka, S., and Nitta, T. 2002. Ethanol oxidation reactions catalyzed by water molecules: $CH_3CH_2OH + nH_2O \rightarrow CH_3CHO + H_2 + nH_2O$ ($n = 0, 1, 2$). *Chemical Physics Letters* 363 (1):80–86.
24. Arita, T., Nakahara, K., Nagami, K., and Kajimoto, O. 2003. Hydrogen generation from ethanol in supercritical water without catalyst. *Tetrahedron Letters* 44 (5):1083–1086.
25. Jin, F., Zhou, Z., Enomoto, H., Moriya, T., and Higashijima, H. 2004. Conversion mechanism of cellulosic biomass to lactic acid in subcritical water and acid-base catalytic effect of subcritical water. *Chemistry Letters* 33 (2):126–127.
26. Shen, Z., Zhang, Y., Jin, F., Zhou, X., Kishita, A., and Tohji, K. 2010. Hydrogen-transfer reduction of ketones into corresponding alcohols using formic acid as a hydrogen donor without a metal catalyst in high-temperature water. *Industrial & Engineering Chemistry Research* 49 (13):6255–6259.
27. Horita, J., and Berndt, M. E. 1999. Abiogenic methane formation and isotopic fractionation under hydrothermal conditions. *Science* 285 (5430):1055–1057.
28. Takahashi, H., Kori, T., Onoki, T., Tohji, K., and Yamasaki, N. 2008. Hydrothermal processing of metal based compounds and carbon dioxide for the synthesis of organic compounds. *Journal of Materials Science* 43 (7):2487–2491.
29. Takahashi, H., Liu, L. H., Yashiro, Y., Ioku, K., Bignall, G., Yamasaki, N., and Kori, T. 2006. CO_2 reduction using hydrothermal method for the selective formation of organic compounds. *Journal of Materials Science* 41 (5):1585–1589.
30. Wu, B., Gao, Y., Jin, F., Cao, J., Du, Y., and Zhang, Y. 2009. Catalytic conversion of $NaHCO_3$ into formic acid in mild hydrothermal conditions for CO_2 utilization. *Catalysis Today* 148 (3):405–410.
31. He, C., Tian, G., Liu, Z., and Feng, S. 2010. A mild hydrothermal route to fix carbon dioxide to simple carboxylic acids. *Organic Letters* 12 (4):649–651.
32. Shen, Z., Zhang, Y., and Jin, F. 2011. From $NaHCO_3$ into formate and from isopropanol into acetone: Hydrogen-transfer reduction of $NaHCO_3$ with isopropanol in high-temperature water. *Green Chemistry* 13 (4):820–823.
33. Jin, F., Gao, Y., Jin, Y., Zhang, Y., Cao, J., Wei, Z., and Smith Jr, R. L. 2001. High-yield reduction of carbon dioxide into formic acid by zero-valent metal/metal oxide redox cycles. *Energy & Environmental Science* 4 (3):881–884.

34. Jin, F., Moriya, T., and Enomoto, H. 2003. Oxidation reaction of high molecular weight carboxylic acids in supercritical water. *Environmental Science & Technology* 37 (14):3220–3231.
35. Behr, A., Eilting, J., Irawadi, K., Leschinski, J., and Lindner, F. 2008. Improved utilisation of renewable resources: New important derivatives of glycerol. *Green Chemistry* 10 (1):13–30.
36. Zheng, Y., Chen, X., and Shen, Y. 2008. Commodity chemicals derived from glycerol, an important biorefinery feedstock. *Chemical Reviews* 108 (12):5253–5277.
37. Kishida, H., Jin, F., Zhou, Z., Moriya, T., and Enomoto, H. 2005. Conversion of glycerin into lactic acid by alkaline hydrothermal reaction. *Chemistry Letters* 34 (11):1560–1561.
38. Kishida, H., Jin, F., Moriya, T., and Enomoto, H. 2006. Kinetic study on conversion of glycerin to lactic acid by alkaline hydrothermal reaction. *Kagaku Kogaku Ronbunshu* 32 (6):535–541.
39. Shen, Z., Jin, F., Zhang, Y., Wu, B., Kishita, A., Tohji, K., and Kishida, H. 2009. Effect of alkaline catalysts on hydrothermal conversion of glycerin into lactic acid. *Industrial & Engineering Chemistry Research* 48 (19):8920–8925.
40. Amass, W., Amass, A., and Tighe, B. 1998. A review of biodegradable polymers: Uses, current developments in the synthesis and characterization of biodegradable polyesters, blends of biodegradable polymers and recent advances in biodegradation studies. *Polymer International* 47 (2):89–144.
41. Inoue, Y., Izumida, H., Sasaki, Y., and Hashimoto, H. 1976. Catalytic fixation of carbon dioxide to formic acid by transition-metal complexes under mild conditions. *Chemistry Letters* 5 (8):863–864.
42. Ai, M., and Ohdan, K. 1999. Formation of pyruvaldehyde (2-oxopropanal) by oxidative dehydrogenation of hydroxyacetone. *Bulletin of the Chemical Society of Japan* 72 (9):2143–2148.
43. Jin, F., and Enomoto, H. 2011. Rapid and highly selective conversion of biomass into value-added products in hydrothermal conditions: Chemistry of acid/base-catalysed and oxidation reactions. *Energy & Environmental Science* 4 (2):382–397.
44. Shen, Z., Jin, F., Zhang, Y., Wu, B., and Cao, J. 2009. Hydrogen transfer reduction of ketones using formic acid as a hydrogen donor under hydrothermal conditions. *Journal of Zhejiang University Science A* 10 (11):1631–1635.
45. Jin, F., Zhang, Y., and Shen, Z. Method for producing aminic acid by glycerin water thermal reduction of CO_2, China, Patent No.: ZL200810039421.X. European application No.: CN20081039421 20080624.
46. Lokras, S. S., Deshpande, P. K., and Kuloor, N. R. 1970. Catalytic dehydrogenation of 2-propanol to acetone. *Industrial & Engineering Chemistry Process Design and Development* 9 (2):293–297.
47. Tsai, J. C., and Nicholas, K. M. 1992. Rhodium-catalyzed hydrogenation of carbon dioxide to formic acid. *Journal of the American Chemical Society* 114 (13):5117–5124.
48. Tho, N. M., and Ha, T. K. 1984. A theoretical study of the formation of carbonic acid from the hydration of carbon dioxide: A case of active solvent catalysis. *Journal of the American Chemical Society* 106 (3):599–602.
49. Chheda, J .N., Huber, G. W., and Dumesic, J. A. 2007. Liquid-phase catalytic processing of biomass-derived oxygenated hydrocarbons to fuels and chemicals. *Angewandte Chemie International Edition* 46 (38):7164–7183.
50. Corma, A., Iborra, S., and Velty, A. 2007. Chemical routes for the transformation of biomass into chemicals. *Chemical Reviews* 107 (6):2411–2502.
51. Shen, Z., Zhang, Y., and Jin, F. 2012. The alcohol-mediated reduction of CO_2 and $NaHCO_3$ into formate: A hydrogen transfer reduction of $NaHCO_3$ with glycerine under alkaline hydrothermal conditions. *RSC Advances* 2 (3):797–801.

52. Zhang, Y., Shen, Z., Zhou, X., Zhang, M., and Jin, F. 2012. Solvent isotope effect and mechanism for the production of hydrogen and lactic acid from glycerol under hydrothermal alkaline conditions. *Green Chemistry* 14 (12):3285–3288.
53. Lira, C. T., and McCrackin, P. J. 1993. Conversion of lactic acid to acrylic acid in near-critical water. *Industrial & Engineering Chemistry Research* 32 (11):2608–2613.
54. Mok, W. S. L., Antal Jr, M. J., and Jones Jr, M. 1989. Formation of acrylic acid from lactic acid in supercritical water. *The Journal of Organic Chemistry* 54 (19):4596–4602.
55. Kuhlmann, B., Arnett, E. M., and Siskin, M. 1994. Classical organic reactions in pure superheated water. *The Journal of Organic Chemistry* 59 (11):3098–3101.
56. Kuhlmann, B., Arnett, E. M., and Siskin, M. 1994. HD exchange in pinacolone by deuterium oxide at high temperature and pressure. *The Journal of Organic Chemistry* 59 (18):5377–5380.
57. Bagnall, L., and Strauss, C. R. 1999. Uncatalysed hydrogen-transfer reductions of aldehydes and ketones. *Chemical Communications* (3):287–288.
58. Lermontov, S. A., Shkavrov, S. V., and Kuryleva, N. V. 2003. Uncatalyzed Meerwein–Ponndorf–Verley reduction of trifluoromethyl carbonyl compounds by high-temperature secondary alcohols. *Journal of Fluorine Chemistry* 121 (2):223–225.
59. Sominsky, L., Rozental, E., Gottlieb, H., Gedanken, A., and Hoz, S. 2004. Uncatalyzed Meerwein–Ponndorf–Oppenauer–Verley reduction of aldehydes and ketones under supercritical conditions. *The Journal of Organic Chemistry* 69 (5):1492–1496.
60. Knill, C. J., and Kennedy, J. F. 2003. Degradation of cellulose under alkaline conditions. *Carbohydrate Polymers* 51 (3):281–300.
61. Ulgen, A., and Hoelderich, W. 2009. Conversion of glycerol to acrolein in the presence of WO_3/ZrO_2 catalysts. *Catalysis Letters* 131 (1–2):122–128.
62. Vasconcelos, S. J. S., Lima, C. L., Filho, J. M., Oliveira, A. C., Barros, E. B., de Sousa, F. F., Rocha, M. G. C., Bargiela, P., and Oliveira, A. C. 2011. Activity of nanocasted oxides for gas-phase dehydration of glycerol. *Chemical Engineering Journal* 168 (2):656–664.

12 Hydrothermal Reduction of CO_2 with Compounds Containing Nitrogen

Guodong Yao, Feiyan Chen, Jia Duo, Fangming Jin, and Heng Zhong

CONTENTS

12.1 Introduction 185
12.2 Experimental Section 187
    12.2.1 Materials 187
    12.2.2 Experimental Procedure 187
    12.2.3 Product Analysis 187
12.3 Results and Discussion 187
    12.3.1 Possibility of Reduction of $NaHCO_3$ by $N_2H_4 \cdot H_2O$ 187
    12.3.2 Effects of Reaction Parameters on the Yield of Formate with Ni Catalyst 189
    12.3.3 Effects of Reaction Parameters on the Yield of Formate with Ni and ZnO Catalysts 192
    12.3.4 Optimal Reaction Parameters of the Experiment with Ni and ZnO Catalysts 192
    12.3.5 Proposed Mechanism of CO_2 Reduction with $N_2H_4 \cdot H_2O$ 194
12.4 Conclusions 196
Acknowledgments 196
References 196

12.1 INTRODUCTION

The increase of atmospheric CO_2 linked to global warming is threatening the Earth's environment and sustainable development. Therefore, reduction of CO_2 emissions is urgently needed. Hydrogenation of CO_2 into chemicals or fuels could be regarded as an efficient way to reduce CO_2 emissions and to minimize global warming.[1,2] In the past several decades, many promising methods for CO_2 reduction have been proposed.[3–5] Among these methods, the photochemical reduction of CO_2 is regarded as the most promising approach, but its efficiency is low and it has limited applications on a large scale. Moreover, the elaborately prepared catalysts are commonly required.[6] The hydrogenation of CO_2 with gaseous hydrogen is currently considered to be the most commercially feasible synthetic route. However, high-pressure and high-purity hydrogen used in this method solves storage and transportation problems. Moreover, because of the thermodynamical stability of gaseous hydrogen and CO_2, the hydrogenation of CO_2 is always associated with the finding or synthesis of highly reactive catalysts, mainly involving noble metal (Ir, Ru, Rh, etc.) complexes.[7-10] Therefore, developing an efficient method for the hydrogenation of CO_2 with a new hydrogen source and a simple catalyst is urgently required.

Some inorganic and organic compounds having nitrogen that possesses reductive amine groups exhibit satisfactory reductive ability under hydrothermal conditions. Moreover, these compounds are actually considered nonflammable when being transported, whereas hydrogen burns with an invisible flame. Therefore, these compounds containing amine groups are safe and simple hydrogen sources. Hydrous hydrazine is a typical compound having an amine group. It contains 7.9% hydrogen and is therefore proposed as a promising indirect liquid hydrogen storage material.[11] Recently, several studies on hydrogen production by the catalytic decomposition of $N_2H_4 \cdot H_2O$ have been reported.[12,13] However, to the best of our knowledge, there has been no report of CO_2 reduction with $N_2H_4 \cdot H_2O$ as an in situ liquid hydrogen source.

Hydrothermal reactions have been attracting increased attention in chemical synthesis because high-temperature water (HTW) is an environmentally benign solvent compared to organic solvents. Moreover, HTW has unique inherent properties such as a high ion product (K_w) and fewer and weaker hydrogen bonds.[14-16] Thus, some reactions that hardly proceed at low temperatures can occur in HTW.[7-19] Recently, there is also an increasing interest in CO_2 reduction under hydrothermal conditions. For example, our group and others have demonstrated the reduction of CO_2 into value-added chemicals using metals as the reductant without expensive or complex catalysts in HTW.[20-23] The results indicate that in situ hydrogen generated from water splitting by metal possesses higher reactivity than gaseous hydrogen. Moreover, HTW facilitates CO_2 activation and improves the reaction rate. Thus, there is a possibility for the efficient reduction of CO_2 into chemicals by using $N_2H_4 \cdot H_2O$ as an in situ hydrogen source in HTW.

Herein, we present a novel method for the hydrogenation of CO_2 into chemicals under hydrothermal conditions by employing $N_2H_4 \cdot H_2O$ as a liquid hydrogen source with a simple and economical catalyst (Scheme 12.1).

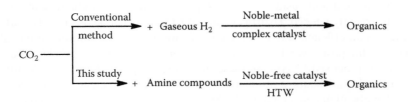

SCHEME 12.1 Proposed method for hydrogenation of CO_2.

Hydrothermal Reduction of CO_2 with Compounds Containing Nitrogen

12.2 EXPERIMENTAL SECTION

12.2.1 Materials

$NaHCO_3$ as the CO_2 source is convenient for experimental operation. Moreover, salts containing HCO_3^- are products formed by carbon capture. Thus, $NaHCO_3$ is representative as a CO_2 source. $N_2H_4 \cdot H_2O$ (85%), $NaHCO_3$ (AR, 98%), Ni, and ZnO powder (200 mesh) were purchased from Sinopharm Chemical Reagent Co., Ltd. Formic acid (AR, 98%) and acetic acid (99.8%) were ordered from Sigma-Aldrich. All reagents were used as test materials without further purification.

12.2.2 Experimental Procedure

The SUS 316 batch reactor and salt bath heater were used in the study and a detailed description can be found elsewhere.[24,25] In a typical procedure, the desired amounts of $N_2H_4 \cdot H_2O$, $NaHCO_3$, Ni, and ZnO with deionized water were loaded in a batch reactor. Subsequently, the sealed reactor was put into a salt bath that had been preheated to the desired temperature. The salt bath can rapidly heat the batch from room temperature to 300°C within 20 s. After the reaction, the reactor was removed from the salt bath to quench in a cold-water bath. When the reactor cooled down to room temperature, the valve was opened and the gas sample was collected using a graduated cylinder in the tank that was filled with saturated NaCl solution. The liquid sample was collected and filtered with a 0.45-μm filter membrane. The solid sample was washed with distilled water and absolute ethanol several times and dried in air for analysis.

12.2.3 Product Analysis

After the reactions, the liquid, gas, and solid samples were respectively collected for analysis. Liquid samples were analyzed by GC-MS, HPLC, and TOC. Gas samples were analyzed by TCD, and solid residues were measured by XRD. Quantitative analyses of formic acid and acetic acid were based on the average value obtained from two sample analyses with the relative errors always less than 5% for all experiments. The HPLC measurement parameters have been described in detail elsewhere.[24] Considering that the real product should be formate because of the alkalinity of $N_2H_4 \cdot H_2O$, the formate yield, which was defined as the percentage of formate to the initial amount of $NaHCO_3$ based on carbon, was used to assess $NaHCO_3$ reduction in this study. The selectivity of formic acid is defined as the percentage of carbon contained in formic acid in relation to the total organic carbon in the liquid phase.

12.3 RESULTS AND DISCUSSION

12.3.1 Possibility of Reduction of $NaHCO_3$ by $N_2H_4 \cdot H_2O$

To investigate the possibility of CO_2 reduction into chemicals with $N_2H_4 \cdot H_2O$, experiments with $N_2H_4 \cdot H_2O$ and $NaHCO_3$ were conducted at 300°C. This temperature was selected because our previous studies have shown that 300°C was the optimum temperature for NaHCO$_3$ reduction into chemicals.[26,27] Analyses of the liquid samples with GC-MS and HPLC showed that the major product was formic acid (Figures 12.1 and 12.2). As shown in Table 12.1, no formic acid and 18% yield of formic acid were obtained in the absence and presence of N$_2$H$_4$·H$_2$O (entries 1 and 2), respectively. These results indicate that NaHCO$_3$ can be selectively reduced into formic acid by N$_2$H$_4$·H$_2$O.

Although NaHCO$_3$ can be reduced into formic acid without the addition of a catalyst, the formate yield was low. Thus, various metals were examined for screening an active catalyst to enhance the yield of formate. As shown in Table 12.1, among the metals investigated, only in the case where Ni was used did the yield of formate increase (entry 4). XRD analyses (Figure 12.3) of the solid residues after the reactions showed that Ni was still present in the zero-valent form of the metal, suggesting that Ni acts as a catalyst in the reduction of NaHCO$_3$ into formate with N$_2$H$_4$·H$_2$O.

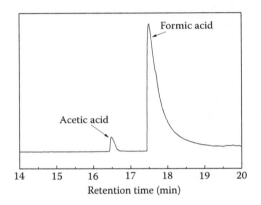

FIGURE 12.1 GC-MS chromatogram of liquid products.

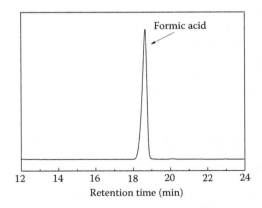

FIGURE 12.2 HPLC chromatogram of liquid products.

TABLE 12.1
The Yields of Formate from the Reduction of NaHCO$_3$ by N$_2$H$_4$·H$_2$O with Different Catalysts[a]

Entry	Reductant	Catalyst	Yield[a]
1	–	–	0
2	N$_2$H$_4$ H$_2$O	–	18
3	–	Ni	1
4	N$_2$H$_4$ H$_2$O	Ni	23
5	N$_2$H$_4$ H$_2$O	Cu	9
6	N$_2$H$_4$ H$_2$O	Fe	13
7	N$_2$H$_4$ H$_2$O	Co	3
8	N$_2$H$_4$ H$_2$O	AlNi	4
9	N$_2$H$_4$ H$_2$O	WC	6

Reaction conditions: NaHCO$_3$: 0.5 M; N$_2$H$_4$ H$_2$O: 2 M; Ni: 4 mmol; water filling: 35%; time: 120 min; temperature 300°C.

[a] The formate yield is defined as the percentage of formate to initial NaHCO$_3$ on the carbon basis.

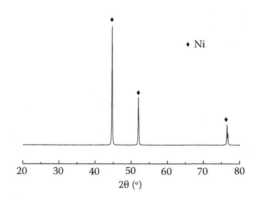

FIGURE 12.3 XRD patterns of solid products after the reaction (NaHCO$_3$: 0.5 M; N$_2$H$_4$·H$_2$O: 2 M; Ni: 4 mmol; water filling: 35%; time: 120 min; temperature: 300°C).

12.3.2 Effects of Reaction Parameters on the Yield of Formate with Ni Catalyst

To obtain the optimal reaction conditions of formate production, a series of experiments were conducted over a wide range of conditions by changing the concentration of reactants, water filling, reaction temperature, and reaction time. First, the effect of the catalyst amount was studied. As shown in Figure 12.4, the yield of formate increased to 27% with 2 mmol Ni. However, formate yield decreased when

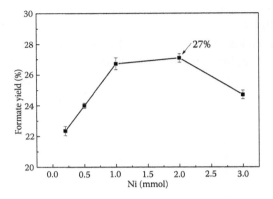

FIGURE 12.4 Effect of the amount of Ni on the yields of formate ($NaHCO_3$: 0.5 M; $N_2H_4 \cdot H_2O$: 2 M; water filling: 35%; time: 120 min; temperature: 300°C).

the amount of Ni further increased to 3 mmol. The reason for this decrease in formate yield with over 2 mmol of Ni is most likely the conversion of formate into CH_4, which was confirmed by TCD analysis (wherein CH_4 was detected with 3 mmol of Ni). Ni also promotes the conversion of formate into CH_4.

Generally, the pressure of the reaction reactor should have a strong effect on the yield of formate. The effect of the pressure was investigated by varying water filling in the batch reactor. As shown in Figure 12.5a, the increase in water filling/pressure is favorable for $NaHCO_3$ reduction into formate, and the yield of formate significantly increased to nearly 50% as the water filling increased to 60%. A possible reason for the increase in formate yield with an increase of water filling could be explained by the fact that most of the hydrogen that is formed in situ is being dissolved in water under a higher pressure, which facilitates the reaction of H_2 with $NaHCO_3$. Considering that the yield of formate acid with 55% and 60% water filling was not significantly different, the following experiments were conducted with 55% water filling. A higher amount of the reductant $N_2H_4 \cdot H_2O$ should also be favorable for $NaHCO_3$ reduction but it is costly. Thus, to obtain an optimized amount of $N_2H_4 \cdot H_2O$ (or ratio of $N_2H_4 \cdot H_2O$ to $NaHCO_3$), the effect of the amount of $N_2H_4 \cdot H_2O$ was investigated with a fixed amount of $NaHCO_3$ (0.5 M). As shown in Figure 12.5a, the yield of formate greatly increased with an increase in the concentration of $N_2H_4 \cdot H_2O$; however, as the concentration of $N_2H_4 \cdot H_2O$ further increased to 6 M, the formate yield remained stable.

Then the effects of reaction time and temperature on the yield of formate were further investigated under the optimal conditions obtained above. As shown in Figure 12.5b, the formate yield increased first and then decreased with an increase in the reaction time, and the highest yield was obtained after 60 min. The yield of formate should be related to the balance of its formation and decomposition during the reactions. The decrease in the formate yield after 60 min indicates that the rate of decomposition of formate was faster than that of the formation. For the effect of temperature, as shown in Figure 12.5b, the formate yield drastically increased as the reaction temperature increased from 250°C to 300°C and reached a maximum at 300°C. The observed decrease in the formate yield at temperatures above 300°C could also be caused by the decomposition of the formate.

Hydrothermal Reduction of CO_2 with Compounds Containing Nitrogen

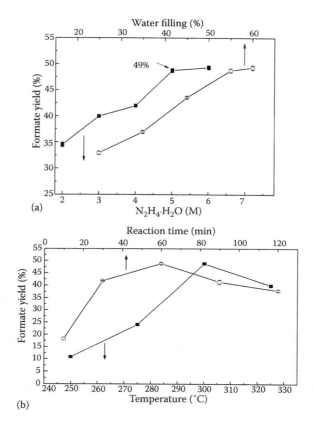

FIGURE 12.5 (a) Effects of water filling and the amount of $N_2H_4 \cdot H_2O$ on the yields of formate (time: 60 min; temperature: 300°C). (b) Effects of reaction time and temperature on the yields of formate ($N_2H_4 \cdot H_2O$: 5 M; water filling: 55%).

The organic carbon in the liquid samples was also determined by TOC analysis, and the amount of carbon in the formate was comparable to the total carbon in the samples, which indicates that the product of the hydrogenation of $NaHCO_3$ was almost all formate. The main by-product was acetic acid with a yield of 0.4%, and, thus, the selectivity for the production of formate was approximately 99%.

Finally, the stability of the Ni catalyst was examined. As shown in Figure 12.6, although a slight decrease in the yield of formate was observed in the second cycle, a further decrease was not observed, indicating that Ni can remain stable after two cycles. The specific surface area of the Ni powder was measured by BET, which showed that the specific surface area of Ni decreased from 2290 m²/kg before the reaction to 2196 m²/kg after the first reaction; this phenomenon may partly explain the decrease in the formate yield. Moreover, to assess the leaching of the Ni catalyst and the main compositions (Fe, Mn) of the SUS316 reactor wall, the ion concentration of Ni, as well as Fe and Mn, in the solution after the reactions was measured by ICP; a small of amount of these ions was found in the liquid sample with Ni at

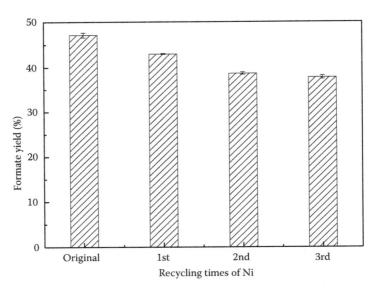

FIGURE 12.6 Effect of reused Ni on the yields of formate (NaHCO$_3$: 0.5 M; N$_2$H$_4$·H$_2$O: 5 M; Ni: 3.135 mmol; water filling: 55%; time: 60 min; temperature: 300°C).

2.126 parts per million (ppm), Fe at 0.5366 ppm, and Mn at 0.1642 ppm. This outcome indicates that the leaching of these metals was very low. To further examine the possible catalytic activity of Fe or Mn from the reactor wall, the formate yield with and without the addition of Fe or Mn ions was compared. A 10^4-fold concentration of Fe or Mn ions compared to the detected concentration of Fe or Mn without the addition of Fe or Mn was added. The results showed that the formate yields with Fe or Mn had no obvious change compared to that without the addition of Fe or Mn. Thus, the reactor walls had no significant catalytic role in the reduction of NaHCO$_3$.

12.3.3 Effects of Reaction Parameters on the Yield of Formate with Ni and ZnO Catalysts

ZnO is a traditional hydrogenation catalyst. To further improve conversion efficiency, ZnO was employed as a co-catalyst in the catalytic reduction of CO$_2$ with hydrazine under hydrothermal conditions. The result showed that a 55.5% yield of formic acid was achieved with Ni and ZnO catalysts, whereas a 49% formate yield was obtained only with Ni. XRD analyses of solid samples (Figure 12.7) showed that the phase states of Ni and ZnO were not changed before and after the reaction, indicating that Ni and ZnO acted as co-catalysts in the hydrogenation of NaHCO$_3$.

12.3.4 Optimal Reaction Parameters of the Experiment with Ni and ZnO Catalysts

To obtain the optimal reaction conditions, a series of experiments were conducted where the amount of ZnO, N$_2$H$_4$·H$_2$O concentration, water filling, reaction time, and

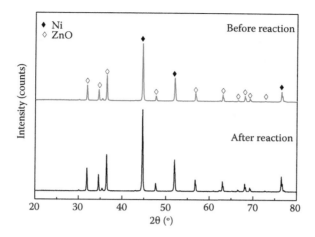

FIGURE 12.7 XRD patterns of solid products before and after the reaction ($NaHCO_3$: 0.5 M; $N_2H_4 \cdot H_2O$: 5 M; Ni: 3.14 mmol; ZnO: 1.25 mmol; water filling: 35%; time: 60 min; temperature: 300°C).

temperature are varied. A previous study showed that 3.14 mmol Ni is suitable for the catalytic reduction of 0.5 M $NaHCO_3$ into formic acid, and therefore, we only changed the ZnO amount from 0 to 1.57 mmol in this study. As shown in Figure 12.8, the formic acid yield increased to 61.5% with 1.25 mmol ZnO. However, the yield of formic acid decreased to 60.3% with 1.57 mmol ZnO. This yield change is probably attributed to the decomposition of formic acid into methane, which was supported

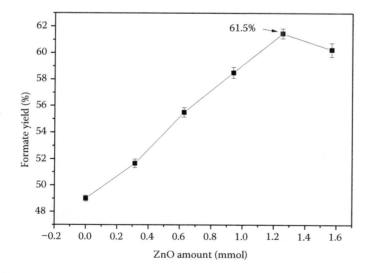

FIGURE 12.8 Effect of the amount of ZnO on the yield of formic acid ($NaHCO_3$: 0.5 M; $N_2H_4 \cdot H_2O$: 5 M; Ni: 3.135 mmol; water filling: 55%; time: 60 min; temperature: 300°C).

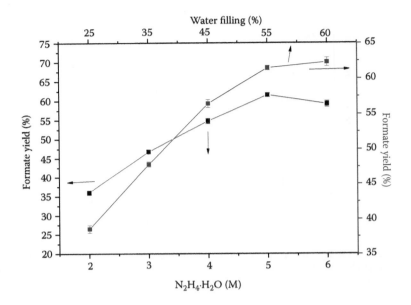

FIGURE 12.9 Effects of the amount of $N_2H_4 \cdot H_2O$ and water filling on the yield of formic acid ($NaHCO_3$: 0.5 M; Ni: 3.14 mmol; ZnO: 1.25 mmol; time: 60 min; temperature: 300°C).

by our previous study.[23] Thus, the amount of ZnO was 1.25 mmol in the following experiments.

Subsequently, the influence of the $N_2H_4 \cdot H_2O$ concentration on the formic acid yield was studied. As shown in Figure 12.9, the yield of formic acid obviously increased with an increase in the concentration of $N_2H_4 \cdot H_2O$; however, as the concentration of $N_2H_4 \cdot H_2O$ further increased to 6 M, the formic acid yield changed slightly (62.3%). Thus, 5 M $N_2H_4 \cdot H_2O$ was used as the reductant in subsequent experiments. Then the influence of the pressure on the formic acid yield was investigated by changing water filling in a batch reactor. As shown in Figure 12.9, the increase in water filling is favorable for $NaHCO_3$ reduction into formic acid, and the yield of formic acid remarkably increased to 61.5% as the water filling increased to 55%. The reason could be the same as that in the experiment with Ni. The effects of reaction time and temperature on the yield of formic acid were also studied under the optimal conditions obtained above. As shown in Figure 12.10, the maximum 61.5% yield was obtained after 60 min at 300°C.

12.3.5 Proposed Mechanism of CO_2 Reduction with $N_2H_4 \cdot H_2O$

In several recent reports, Raney Ni or bimetallic nanoparticle containing Ni can catalytically decompose $N_2H_4 \cdot H_2O$ to hydrogen with high selectivity (Scheme 12.2).[12,13] In this study, bulk Ni powder not only can promote the decomposition of $N_2H_4 \cdot H_2O$ to hydrogen, which was examined in our previous study, but also can catalyze hydrogenation of $NaHCO_3$. Although ZnO cannot promote hydrogen production from

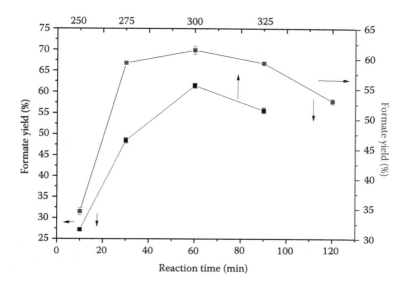

FIGURE 12.10 Effects of reaction time and temperature on the yield of formic acid (NaHCO$_3$: 0.5 M; N$_2$H$_4$·H$_2$O: 5 M; Ni: 3.14 mmol; ZnO: 1.25 mmol; water filling: 55%).

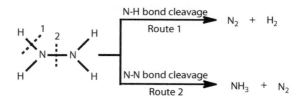

SCHEME 12.2 Decomposition ways of N$_2$H$_4$·H$_2$O.

N$_2$H$_4$·H$_2$O, it can act as a co-catalyst to accelerate NaHCO$_3$ hydrogenation. To investigate the difference between the activity of the in situ hydrogen from N$_2$H$_4$·H$_2$O and gaseous hydrogen, an experiment with gaseous hydrogen as the hydrogen source instead of N$_2$H$_4$·H$_2$O was conducted under optimal conditions (NaHCO$_3$: 0.5 M; Ni: 3.14 mmol; ZnO: 1.25 mmol; time: 60 min; water filling: 55%; temperature: 300°C). We injected 2.5 MPa gaseous hydrogen, which is approximately equal to the amount of hydrogen produced from the experiment with N$_2$H$_4$·H$_2$O under optimal conditions, into the reactor. Only a 2.6% yield of formic acid was obtained. Although further evidence is needed, the present results suggest that the in situ hydrogen formed from N$_2$H$_4$·H$_2$O may have a higher activity than the gaseous hydrogen in NaHCO$_3$ reduction into formic acid.

Based on the experimental results, a possible mechanism was proposed as shown in Scheme 12.3. Initially, the H proton of N$_2$H$_4$·H$_2$O is adsorbed onto the Ni surface and N$_2$H$_4$·H$_2$O prefers the cleavage of the N–H bond, leading to H$_2$ formation. Subsequently, the in situ generated H$_2$ is activated on Ni and ZnO surfaces. Then, the active H attacks the carbon of C=O, followed by the hydroxyl group of HCO$_3^-$

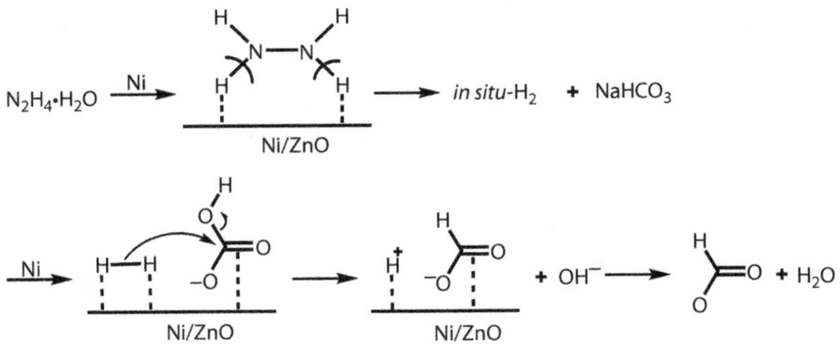

SCHEME 12.3 Proposed mechanism of reduction of $NaHCO_3$ into formic acid with $N_2H_4 \cdot H_2O$ over Ni and ZnO.

leaving. Finally, formate was obtained together with H_2O, which was formed by the combination of another active H of H_2 and the hydroxyl group of HCO_3^- leaving.

12.4 CONCLUSIONS

A novel efficient method for the selective reduction of $NaHCO_3$ into formic acid by using $N_2H_4 \cdot H_2O$ over a common Ni and ZnO powder in HTW was proposed. Nearly 99% selectivity and approximately 62% yield of formic acid were achieved. The results indicate that $N_2H_4 \cdot H_2O$ is an excellent in situ hydrogen source for $NaHCO_3$ reduction. The common Ni and ZnO powders exhibit satisfactory catalytic activity and stability in HTW. In addition, we have found that CO_2 could be reduced to CH_4 by adjusting pH and reaction time with $N_2H_4 \cdot H_2O$. Further work is now in progress.

The proposed method is simple and safe because it requires neither elaborately prepared catalysts nor high-purity hydrogen and hydrogen storage. This approach is valuable for the conversion of CO_2 into value-added chemicals.

ACKNOWLEDGMENTS

The authors thank the financial support of the National Natural Science Foundation of China (Nos. 21277091 and 51472159), the State Key Program of National Natural Science Foundation of China (No. 21436007), Key Basic Research Projects of Science and Technology Commission of Shanghai (No. 14JC1403100), and China Postdoctoral Science Foundation (No. 2013 M541520).

REFERENCES

1. T. Sakakura, J. C. Choi, H. Yasuda. Transformation of carbon dioxide. *Chemical Reviews*, *107*(6): 2365–2387, **2007**.
2. K. M. K. Yu, I. Curcic, J. Gabriel, S. C. E. Tsang. Recent advances in CO_2 capture and utilization. *Chemsuschem*, *1*(11): 893–899, **2008**.
3. J. C. Tsai, K. M. Nicholas. Rhodium-catalyzed hydrogenation of carbon-dioxide to formic-acid. *Journal of the American Chemical Society*, *114*(13): 5117–5124, **1992**.

4. M. Aresta, A. Dibenedetto. Utilisation of CO_2 as a chemical feedstock: Opportunities and challenges. *Dalton Transactions*, (28): 2975–2992, **2007**.
5. H. Takeda, H. Koizumi, K. Okamoto, O. Ishitani. Photocatalytic CO_2 reduction using a Mn complex as a catalyst. *Chemical Communications*, 50(12): 1491–1493, **2014**.
6. G. Centi, S. Perathoner. Opportunities and prospects in the chemical recycling of carbon dioxide to fuels. *Catalysis Today*, 148(3–4): 191–205, **2009**.
7. P. G. Jessop, Y. Hsiao, T. Ikariya, R. Noyori. Homogeneous catalysis in supercritical fluids: Hydrogenation of supercritical carbon dioxide to formic acid, alkyl formates, and formamides. *Journal of the American Chemical Society*, 118(2): 344–355, **1996**.
8. C. C. Tai, J. Pitts, J. C. Linehan, A. D. Main, P. Munshi, P. G. Jessop. In situ formation of ruthenium catalysts for the homogeneous hydrogenation of carbon dioxide. *Inorganic Chemistry*, 41(6): 1606–1614, **2002**.
9. P. Munshi, A. D. Main, J. C. Linehan, C. C. Tai, P. G. Jessop. Hydrogenation of carbon dioxide catalyzed by ruthenium trimethylphosphine complexes: The accelerating effect of certain alcohols and amines. *Journal of the American Chemical Society*, 124(27): 7963–7971, **2002**.
10. X. Zhou, J. Qu, F. Xu, J. Hu, J. S. Foord, Z. Zeng, X. Hong, S. C. E. Tsang. Shape selective plate-form Ga_2O_3 with strong metal-support interaction to overlying Pd for hydrogenation of CO_2 to CH_3OH. *Chemical Communications*, 49(17): 1747–1749, **2013**.
11. R. Lan, J. T. S. Irvine, S. W. Tao. Ammonia and related chemicals as potential indirect hydrogen storage materials. *International Journal of Hydrogen Energy*, 37(2): 1482–1494, **2012**.
12. L. He, Y. Huang, X. Y. Liu, L. Li, A. Wang, X. Wang, C.-Y. Mou, T. Zhang. Structural and catalytic properties of supported Ni–Ir alloy catalysts for H_2 generation via hydrous hydrazine decomposition. *Applied Catalysis B—Environmental*, 147: 779–788, **2014**.
13. L. He, Y. Huang, A. Wang, X. Wang, X. Chen, J. J. Delgado, T. Zhang. A noble-metal-free catalyst derived from Ni–Al hydrotalcite for hydrogen generation from $N_2H_4 \cdot H_2O$ decomposition. *Angewandte Chemie-International Edition*, 51(25): 6191–6194, **2012**.
14. T. I. Mizan, P. E. Savage, R. M. Ziff. Temperature dependence of hydrogen bonding in supercritical water. *Journal of Physical Chemistry*, 100(1): 403–408, **1996**.
15. P. E. Savage, N. Akiya. Roles of water for chemical reactions in high-temperature water. *Chemical Reviews*, 102(8): 2725–2750, **2002**.
16. P. E. Savage. A perspective on catalysis in sub- and supercritical water. *Journal of Supercritical Fluids*, 47(3): 407–414, **2009**.
17. A. Kruse, E. Dinjus. Hot compressed water as reaction medium and reactant— Properties and synthesis reactions. *Journal of Supercritical Fluids*, 39(3): 362–380, **2007**.
18. M. Sasaki, T. Adschiri, K. Arai. Kinetics of cellulose conversion at 25 MPa in sub- and supercritical water. *AIChE Journal*, 50(1): 192–202, **2004**.
19. N. Akiya, P. E. Savage. Kinetics and mechanism of cyclohexanol dehydration in high-temperature water. *Industrial & Engineering Chemistry Research*, 40(8): 1822–1831, **2001**.
20. J. Duo, F. Jin, Y. Wang, H. Zhong, L. Lyu, G. Yao, Z. Huo. $NaHCO_3$-enhanced hydrogen production from water with Fe and in situ highly efficient and autocatalytic $NaHCO_3$ reduction into formic acid. *Chemical Communications*, 52(16): 3316–3319, **2016**.
21. H. Zhong, Y. Gao, G. Yao, X. Zeng, Q. Li, Z. Huo, F. Jin. Highly efficient water splitting and carbon dioxide reduction into formic acid with iron and copper powder. *Chemical Engineering Journal*, 280: 215–221, **2015**.
22. L. Lyu, X. Zeng, J. Yun, F. Wei, F. Jin. No catalyst addition and highly efficient dissociation of H_2O for the reduction of CO_2 to formic acid with Mn. *Environmental Science & Technology*, 48(10): 6003–6009, **2014**.

23. F. Jin, X. Zeng, J. Liu, Y. Jin, L. Wang, H. Zhong, G. Yao, Z. Huo. Highly efficient and autocatalytic H_2O dissociation for CO_2 reduction into formic acid with zinc. *Scientific Reports*, *4*: 4503, **2014**.
24. F. Jin, Z. Zhou, T. Moriya, H. Kishida, H. Higashijima, H. Enomoto. Controlling hydrothermal reaction pathways to improve acetic acid production from carbohydrate biomass. *Environmental Science & Technology*, *39*(6): 1893–1902, **2005**.
25. F. Jin, T. Moriya, H. Enomoto. Oxidation reaction of high molecular weight carboxylic acids in supercritical water. *Environmental Science & Technology*, *37*(14): 3220–3231, **2003**.
26. F. Jin, Y. Gao, Y. Jin, Y. Zhang, J. Cao, Z. Wei, R. L. Smith. High-yield reduction of carbon dioxide into formic acid by zero-valent metal/metal oxide redox cycles. *Energy & Environmental Science*, *4*(3): 881–884, **2011**.
27. Z. Shen, Y. Zhang, F. Jin. From $NaHCO_3$ into formate and from isopropanol into acetone: Hydrogen-transfer reduction of $NaHCO_3$ with isopropanol in high-temperature water. *Green Chemistry*, *13*(4): 820–823, **2011**.

13 Perspectives and Challenges of CO_2 Hydrothermal Reduction

Ligang Luo, Fangming Jin, and Heng Zhong

CONTENTS

13.1 Introduction 199
13.2 Advantages of CO_2 Hydrothermal Reduction..200
13.3 Proposed Future Research Directions 200
    13.3.1 Simulation for Hydrothermal Vent System.....................................200
    13.3.2 Combination of Hydrothermal Reactions and Photocatalytic Reduction of CO_2 201
    13.3.3 Combination of Hydrothermal and Electrochemical Reduction of CO_2 202
    13.3.4 Developing High Stability of Catalysts 202
13.4 Barriers for CO_2 Hydrothermal Reduction 202
13.5 Conclusions...202
Acknowledgments..203
References...203

13.1 INTRODUCTION

Energy shortage and global warming have become worldwide problems and challenges for humans in the 21st century.[1] One of the causes of the energy crisis and the increase in atmospheric greenhouse gases such as carbon dioxide (CO_2) is the imbalance of the rapid consumption of fossil fuels and the slow rate of fossil fuel formation.[2] The urgent need to reduce atmospheric concentrations of greenhouse gases has prompted action from national and international governments and industries. The conversion of the greenhouse gas CO_2 into value-added carbon chemicals and fuels is a potential method to solve both the energy and environmental problems.[3–5] Currently, the predominant methods proposed to convert CO_2 into chemicals and fuels are hydrogenation catalytic, photocatalytic, and electrochemical methods. However, these processes have disadvantages such as expensive metals, additional hydrogen needed, low efficiency, and slow conversion rate. Thus, developing a highly efficient and simple process for CO_2 reduction is highly desired.

Hydrothermal reactions have played an important role in the formation of fossil fuels, for example, the abiotic conversion of dissolved CO_2 into hydrocarbons and great potential for rapid conversion of biomass into value-added products.[6,7] Recently, some research on the utilization of hydrothermal reactions to CO_2 reduction has demonstrated great potential for the highly efficient and rapid conversion of CO_2 into value-added products.[8–11] Hence, this chapter presents the perspectives and challenges to CO_2 hydrothermal reductions.

13.2 ADVANTAGES OF CO_2 HYDROTHERMAL REDUCTION

The hydrothermal condition has a lot of advantageous properties, such as high self-dissociation constant, weak hydrogen bonds, and polarity, which are favorable to biomass conversion. Thus, many processes on the application of hydrothermal reactions to biomass conversion have been proposed, and hydrothermal conversion of biomass has demonstrated great potential for efficient and quick biomass conversion. Compared with hydrothermal conversion of biomass, although research on the application of hydrothermal reactions to CO_2 reduction is relatively spares, hydrothermal CO_2 reduction is attracting attention from both scientific and engineering standpoints and would be expected to have high potential to quickly, efficiently, and environmentally friendly convert CO_2 into organics. It is because water is an environmentally friendly and recyclable medium, and hydrothermal processes simulate the natural phenomena of abiotic organic synthesis. Some recent advances in hydrothermal CO_2 reduction have shown its advantages, for example, not requiring the use of precious-metal catalysts or harsh reagents, high-purity hydrogen, the in situ use of hydrogen in water for CO_2 reduction, and solving problems regarding hydrogen storage and transformation. Thus, CO_2 hydrothermal reduction can be simple, rapid, and efficient, which are important attributes for industry applications.

13.3 PROPOSED FUTURE RESEARCH DIRECTIONS

13.3.1 Simulation for Hydrothermal Vent System

Since the first discovery of hydrothermal vents in 1977, these deep ocean phenomena have attracted global attention, in which the abiotic synthesis of organic compounds from dissolved inorganic carbon (CO_2, HCO_3^-, and CO_3^{2-}) to organic compounds[12–14] is a central topic. The natural phenomenon of abiotic synthesis of organic compounds at hydrothermal vents provides us a hint on how to highly efficiently achieve CO_2 reduction, and the use of hydrothermal reactions to CO_2 reduction is a new direction for CO_2 reduction.[14,15] Moreover, from the point of view of the Earth's carbon cycle, as shown in Figure 13.1, CO_2 and water react to form a biomass with the help of solar energy, and humans then use this biomass for survival while discharging a large amount of waste. Some wastes are buried under sediment. The enormous heat and pressure in the deep strata turn these organic materials into petroleum, natural gas, or coal. Humans consume these fuels in everyday life, thus again giving off CO_2 and waste. Originally, Earth has an ideal carbon cycle.

Realistically, however, the carbon balance is broken by anthropogenic activities because the imbalance of fossil fuel formation from wastes in the Earth requires hundreds of millions of years, whereas humans use these fossil fuels within 200 to

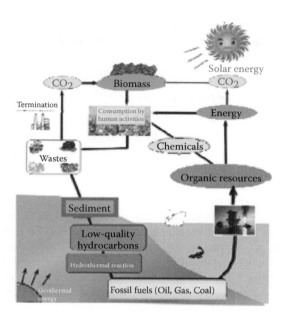

FIGURE 13.1 The pathways for Earth's carbon cycle.

300 years. The length of time required for fossil fuel formation comes from the need for sedimentation of organic wastes to reach the deep underground where the conversion reaction from organic wastes to fuels can occur. Geochemists have experimentally proven that the conversion of organic wastes into oily materials can be sufficiently fast under hydrothermal conditions. Thus, if humans could simulate the natural phenomena of the formation of fossils, then this process should quickly turn not only biomass but also CO_2 into fuels and chemicals. Along this way, CO_2 hydrothermal reduction by mimicking natural abiotic organic synthesis is crucial not only for CO_2 conversion but also for the improvement of the Earth's carbon cycle.

13.3.2 Combination of Hydrothermal Reactions and Photocatalytic Reduction of CO_2

Solar energy, the ultimate energy resource, is the largest exploitable alternative to fossil fuels with enormous potential.[16,17] Artificial photosynthesis, in which solar energy is converted into chemical energy for renewable, non-polluting fuels and chemicals, is regarded as one of the most promising methods for solar energy technologies. However, the direct conversion of solar energy into chemical energy has many problems, such as low conversion efficiencies and low product selectivity. Thus, developing an efficient solar-to-fuel conversion process is a great and fascinating challenge. For this, an integrated system should be expected to improve artificial photosynthesis efficiency and rate. An alternative and probably attractive approach is to combine artificial photosynthesis with hydrothermal reduction, in which the solar energy is used to drive thermodynamic uphill reaction for CO_2 hydrothermal reduction to generate high-value carbon chemicals. In this way, a highly energy-efficient conversion of the solar to fuel process will be performed, and then it is expected to provide a possible avenue for high and rapid artificial photosynthesis.

13.3.3 COMBINATION OF HYDROTHERMAL AND ELECTROCHEMICAL REDUCTION OF CO_2

Electrochemical reduction of CO_2 is also one of the promising methods for synthesizing carbon-containing fuels and energy conversion.[17] Electrochemical reduction of CO_2 in aqueous solutions has been studied for many decades.[18] However, challenges faced by the electrochemical reduction of CO_2 remain, such as low catalytic activity. Furthermore, the major obstacle for electrochemical CO_2 conversion in aqueous solution is the high overpotential required to achieve ideal activity and selectivity.[19,20] These high overpotential issues may be alleviated to some degree by high-temperature water because of its properties, including the dielectric constant of high-temperature water being about an order of magnitude lower than that of water under normal conditions[6] and the specific conductance of high-temperature water being several orders of magnitude higher than that of liquid water.[21,22] Hydrothermal reduction combined with electrochemical reduction methods seems to be a potential method for promoting CO_2 reduction.

13.3.4 DEVELOPING HIGH STABILITY OF CATALYSTS

As mentioned before, among the various options to reduce CO_2 atmospheric loading, hydrothermal reactions have shown a high potential for rapidly and effectively converting CO_2 into useful chemicals. In some cases, without any addition of catalysts, a highly effective conversion of CO_2 into organics can be reached, and also some catalysts for CO_2 hydrothermal reduction have been proposed.[8] However, efficient catalysts, particularly in regard to their stability, are and will be a key factor for rapid and effective CO_2 hydrothermal reduction. Developing an efficient and stable catalytic system under hydrothermal conditions will be an important task in the future.

13.4 BARRIERS FOR CO_2 HYDROTHERMAL REDUCTION

There remain challenges for the extensive practical applications of CO_2 hydrothermal reduction: (1) development of a mild reaction system, (2) development of the reactor with the characteristic of resisting corrosion, and (3) development of highly efficient and stable catalysts. Solving these challenges in the near future will help realize the practical applications of CO_2 hydrothermal reduction.

13.5 CONCLUSIONS

In summary, hydrothermal CO_2 reduction is confirmed to efficiently and rapidly convert CO_2 into high-valued chemicals and fuels. In addition, more intensive work is necessary for the development of hydrothermal CO_2 reductions, such as how to simulate hydrothermal vent systems to form high-valued carbon fuels from CO_2, combining electrochemical and photocatalytic reductions, and developing efficient and stable catalysts, which are key for economically feasible industry applications.

ACKNOWLEDGMENTS

The authors thank the financial support of the National Natural Science Foundation of China (Nos. 21277091 and 51472159), the State Key Program of National Natural Science Foundation of China (No. 21436007), Key Basic Research Projects of Science and Technology Commission of Shanghai (No. 14JC1403100), and China Postdoctoral Science Foundation (No. 2013 M541520).

REFERENCES

1. S. Saeidi, N. A. S. Amin, M. R. Rahimpour, Hydrogenation of CO_2 to value-added products—A review and potential future developments, *Journal of CO_2 Utilization*, **2014**, 5, 66–81.
2. S. J. Davis, K. Caldeira, H. D. Matthews, Future CO_2 emissions and climate change from existing energy infrastructure, *Science*, **2010**, 329(5997), 1330–1333.
3. F. Jin, Y. Gao, Y. Jin, Y. Zhang, J. Cao, Z. Wei, R. L. Smith, High-yield reduction of carbon dioxide into formic acid by zero-valent metal/metal oxide redox cycles, *Energy Environmental Science*, **2011**, 4(3), 881–884.
4. B. Wu, Y. Gao, F. Jin, J. Cao, Y. Du, Y. Zhang, Catalytic conversion of $NaHCO_3$ into formic acid in mild hydrothermal conditions for CO_2 utilization, *Catalysis Today*, **2009**, 148(3), 405–410.
5. G. Yao, F. Chen, Z. Huo, F. Jin, Hydrazine as a facile and highly efficient hydrogen source for reduction of $NaHCO_3$ into formic acid over Ni and ZnO catalysts, *International Journal of Hydrogen Energy*, **2016**, 41(21), 9135–9139.
6. P. E. Savage, Organic chemical reactions in supercritical water, *Chemical Reviews*, **1999**, 99(2), 603–622.
7. F. Jin, Z. Xu, J. Liu, Y. Jin, L. Wang, H. Zhong, G. Yao, Z. Huo, Highly efficient and autocatalytic H_2O dissociation for CO_2 reduction into formic acid with zinc, *Scientific Reports*, **2014**, 4(3), 4503.
8. J. Duo, F. Jin, Y. Wang, H. Zhong, L. Lyu, G. Yao, Z. Huo, $NaHCO_3$-enhanced hydrogen production from water with Fe and in situ highly efficient and autocatalytic $NaHCO_3$ reduction into formic acid, *Chemical Communications*, **2016**, 52(16), 3316–3319.
9. G. Tian, H. Yuan, Y. Mu, C. He, S. Feng, Hydrothermal reactions from sodium hydrogen carbonate to phenol, *Organic Letters*, **2007**, 9(10), 2019–2021.
10. T. Wang, D. Ren, Z. Huo, Z. Song, F. Jin, M. Chen, L. Chen, A nanoporous nickel catalyst for selective hydrogenation of carbonates into formic acid in water, *Green Chemistry*, **2017**, 19, 716–721.
11. Y. Chen, Z. Jing, J. Miao, Y. Zhang, J. Fan, Reduction of CO_2 with water splitting hydrogen under subcritical and supercritical hydrothermal conditions, *International Journal of Hydrogen Energy*, **2016**, 41, 9123–9127.
12. J. M. McDermott, J. S. Seewald, C. R. German, S. P. Sylva, Pathways for abiotic organic synthesis at submarine hydrothermal fields, *Proceedings of the National Academy of Sciences*, **2015**, 112(25), 7668–7672.
13. S. L. Miller, J. L. Bada, Submarine hot springs and the origin of life, *Nature*, **1988**, 334(6183), 609–611.

14. M. S. Dodd, D. Papineau, T. Grenne, J. F. Slack, M. Rittner, F. Pirajno, J. O'Neil, C.T. S. Little, Evidence for early life in Earth's oldest hydrothermal vent precipitates, *Nature*, **2017**, 543(7643), 60–64.
15. B. R. T. Simoneit, Aqueous high-temperature and high-pressure organic geochemistry of hydrothermal vent systems, *Geochimica et Cosmochimica Acta*, **1993**, 57(14), 3231–3243.
16. E. V. Kondratenko, G. Mul, J. Baltrusaitis, G. O. Larrazábal, J. P. Ramírez, Status and perspectives of CO_2 conversion into fuels and chemicals by catalytic, photocatalytic and electrocatalytic processes, *Energy Environmental Science*, **2013**, 6(11), 3112–3135.
17. Y. Qu, X. Duan, Progress, challenge and perspective of heterogeneous photocatalysts, *Chemical Society Reviews*, **2013**, 42(7), 2568–2580.
18. Y. C. Hsieh, S. D. Senanayake, Y. Zhang, W. Xu, D. E. Polyansky, Effect of chloride anions on the synthesis and enhanced catalytic activity of silver nanocoral electrodes for CO_2 electroreduction, *ACS Catalysis*, **2015**, 5(9), 5349–5356.
19. C. Costentin, M. Robert, J. M. Saveant, Catalysis of the electrochemical reduction of carbon dioxide, *Chemical Society Reviews*, **2013**, 42, 2423–2436.
20. D. T. Whipple, P. J. A. Kenis, Prospects of CO_2 utilization via direct heterogeneous electrochemical reduction, *The Journal of Physical Chemistry Letters*, **2010**, 1(24), 3451–3458.
21. N. Akiya, P. E. Savage, Roles of water for chemical reactions in high-temperature water. *Chemical Reviews*, **2002**, 102(8), 2725–2750.
22. K. Sue, K. Arai, Specific behavior of acid–base and neutralization reactions in supercritical water, *The Journal of Supercritical Fluids*, **2004**, 28(1), 57–68.

Index

Page numbers followed by f and t indicate figures and tables, respectively.

A

Abiogenic formate production from CO_2; see also Glycerine
    about, 168–169
    alkaline hydrothermal conversion routes of lactate/ intermediates, 177–179
    alkaline role, verification of, 176
    catalysis of H_2O molecules, 173–174, 174t
    D_2O solvent, effect of, 171–172
    formation of abiogenic formate from CO_2, 169–171
    reaction mechanism
        proposition of, 174–175
        testing of, 175–176, 176t
    reactor materials, 173
Abiotic synthesis of organics, 92
Acetic acid
    decarboxylation of, 10
    production, 83, 83f
Acetol, 176t
Acetone, 163, 167
Acid hydrolysis, 11
Acidic/basic catalysis, 26–27; see also Catalytic hydrothermal reactions
Activation energy
    from glucose and HMF, 13
    metal powders with water, 52t
Adsorption of ions, 23
Alcohol dehydration, 10
Alcohol-mediated reduction of CO_2/$NaHCO_3$ into formate; see also Glycerine
    comparative experiments with or without dry ice, 157–159, 158t
    $NaHCO_3$ quantity/NaOH concentration, 159–160
    reaction pathway on hydrogen-transfer reduction of $NaHCO_3$, 161–163
    reaction temperature and time, effects of, 160
    reactor materials, catalysis effect of, 160–161
Alkaline hydrothermal conversion routes of lactate/ intermediates, 177–179
Alkaline role, verification of, 176
Aluminum
    hydrogen generation with, 38–39
    recycling of, 38
Aluminum and formic acid production
    CO_2 reduction by water splitting with Al, 136–137
    experimental procedure, 130–131
    hydrogenation of CO_2 by water splitting with Al, 132–135
    hydrogen production by water splitting with Al, 131–132, 132t
    materials, 130
    overview, 127–130
    product analysis, 131
    quantum chemical calculations, 131
    waste metal as reducing agent for hydrogenation of CO_2, 135, 135t
Aluminum hydrate, 136
Aqueous sulfur species, hydrothermal reactions of (case study), 27–29
Arrhenius equation, 30, 30f
Arrhenius kinetics, 38
Artificial photosynthesis, 201
Artificial photosynthetic conversion of CO_2, 128
Autocatalyst, 99
Autocatalytic hydrothermal CO_2 reduction with manganese, see Manganese and formic acid production
Autocatalytic mechanism, 75

B

Ball milling, 43
Basic catalysis, 26–27; see also Catalytic hydrothermal reactions
Batch reactor system, 130f
Benzilic acid rearrangement, 175, 176
Bicarbonate, 29, 30
Biomass, 168
Boron, hydrogen production with, 43
Boron oxide gasification, 43

C

Carbon dioxide
    emissions, 61
    in improving hydrogen production from water, 118, 119t
    to methane, hydrothermal reactions from, 85–88
    into organic acid, conversion of (case study), 29–31, 30t
    to phenol, hydrothermal reactions from, 79–81

205

Index

to simple carboxylic acids, hydrothermal reactions from, 81–85
Carbon dioxide conversion; *see also* Methanol production
  to methanol over commercial Cu powder
    general information, 143
    product analysis, 143
    synthesis of methanol from CO_2, 144
  into methanol over Cu nanoparticle
    experimental procedure, 149
    methanol formation, 149–150
Carbon-dioxide hydrothermal reduction, challenges, 199–200
  advantages of, 199
  barriers for, 202
  high stability of catalysts, 202
  hydrothermal and electrochemical reduction, combination of, 202
  hydrothermal reactions and photocatalytic reduction, combination, 201–202
  simulation for hydrothermal vent system, 200–201
Carbon dioxide reduction
  with Fe and formic acid production
    autocatalytic mechanism, 75
    catalytic activity of formed Fe_3O_4, 73–75, 74t, 75t
    in situ hydrogen, role of, 72–73, 73t
  with Mn, potential of, 112–114, 114t
  by water splitting with Al, 136–137
Carboxyl acid, 163
Catalysis of H_2O molecules, 173–174, 174t
Catalysts
  high stability of, 202
  in-situ characterization of, 25
Catalytic hydrogenation of CO_2, 143
Catalytic hydrothermal reactions
  acidic/basic and redox catalysis, 26–27
  case study
    conversion of CO_2 into organic acid, 29–31, 30t
    hydrothermal reactions of aqueous sulfur species, 27–29
  in-situ characterization, 25–26
  interfacial chemistry between solid and solution, 23–24
  kinetic model, 24–25
  quantum chemistry calculation, 26
Cellulose
  hydrolysis of, 11f
  under hydrothermal conditions, 11
Conversion, defined, 54t
Copper-catalyzed hydrothermal CO_2 reduction, *see* Methanol production
Copper nanoparticle, catalytic efficiency of, 149
Copper powder, commercial; *see also* Carbon dioxide conversion

conversion of CO_2 to methanol over, 143–147
methanol production over, 147–149, 148t

D

Decarboxylation of formic acid, 115
Dehydration, 13–14, 13f; *see also* Water under high temperature and pressure (WHTP) conditions
Density functional theory (DFT), 101, 129
Diamond anvil cell (DAC), 25
Dielectric constant; *see also* Water under high temperature and pressure (WHTP) conditions
  about, 3–5, 4f
  HTW, 5–6, 5t, 6t
Diethylamine, 173
Dissociation of water, 115–118
D_2O solvent, effect of, 171–172
Double-metal additives, 57
Dry ice, 155, 157–159, 158t

E

Earth's carbon cycle, 200, 201f
Electrochemical reduction of CO_2, 202
Electrospray ionization (ESI), 156
Eley–Rideal mechanism, 24
Endeavor Catalyst Screening System (ECSS), 80, 81
Energy conversion efficiency assessment, 105–106
Energy crisis, causes of, 199
Esterification of triglycerides, 141
Ethanol oxidation reaction, 10
Excitation energy calculations, 26

F

Fe oxidation
  proposed pathway for, 68–69
  reaction conditions on, 65–68, 66t
Fischer–Tropsch process, 110
Formate production
  with $NaHCO_3$/Zn
    reaction characteristics/optimization of reaction parameters, 94–96
    reduction of $NaHCO_3$, 94
    proposed mechanism for, 99–101, 99t
Formic acid
  decarboxylation/dehydration of, 26
  formation of, 41
  high yield of, 115–118
  selectivity, defined, 131
  use of, 62

Index

Formic acid formation
  investigation via HCO_3/gaseous CO_2, 121–124, 122t
  Mn_xO_y in, 119–121
Formic acid production
  with aluminum, *see* Aluminum and formic acid production
  with gaseous CO_2, 97–98, 98t
  with zinc, *see* Zinc and formic acid production
Formic acid yield
  defined, 131
  reaction conditions on; *see also* Hydrothermal CO_2 reduction with iron
    Fe amount and size of Fe powder, 70, 71f
    initial amount of $NaHCO_3$ and water filling on, 69, 70f
    reaction temperature and time on formic acid yield, 72
  reaction time and temperature on, 134f
Fossil fuel formation, 200
Fourier transform infrared (FT-IR) spectroscopy, 94
Free energy, 102, 103, 104, 123

G

Gamma-valerolactone (GVL), 26
Gas chromatography, 53, 55
Gas chromatography/mass spectroscopy (GC/MS), 63, 64f, 79, 93
Gas chromatography/thermal conductivity detector (GC/TCD), 112, 118
Gasification, hydrothermal, 2
GC analysis, 157
GC-MS chromatogram of liquid products, 188f
GC-TCD, 147
Gibbs free energy, 31f, 55, 56f, 93, 114
Global hydrogen consumption per industry, 47–48, 48f; *see also* Hydrothermal water splitting
Glucose–fructose isomerization, 12, 12f
Glycerine
  abiogenic formate production from CO_2
    about, 168–169
    alkaline hydrothermal conversion routes of lactate/ intermediates, 177–179
    alkaline role, verification of, 176
    catalysis of H_2O molecules, 173–174, 174t
    D_2O solvent, effect of, 171–172
    formation of abiogenic formate from CO_2, 169–171
    reaction mechanism, proposition of, 174–175
    reaction mechanism, testing of, 175–176, 176t
    reactor materials, 173
  alcohol-mediated reduction of CO_2/$NaHCO_3$ into formate
    comparative experiments with or without dry ice, 157–159, 158t
    $NaHCO_3$ quantity and NaOH concentration, 159–160
    reaction pathway on hydrogen-transfer reduction of $NaHCO_3$, 161–163
    reaction temperature and time, effects of, 160
    reactor materials, catalysis effect of, 160–161
  experimental procedure, 155–156
  GC analysis, 157
  HPLC analysis, 157
  hydrogen-transfer reduction of $NaHCO_3$ with isopropanol in HTW
    hydrogen-transfer reduction mechanism, reaction pathway on, 167–168
    NaOH concentration, effect of, 165–166
    quantity of $NaHCO_3$, 164–165, 165t
    reaction products from isopropanol/$NaHCO_3$, 163
    reaction temperature, effect of, 163–164
    reaction time, effect of, 166–167
  LC-MS analysis, 156
  materials, 155
  NMR analysis, 156
  overview, 154–155
  product analysis, 156
Glycerol, 29, 30f
Greenhouse effect, 109
Greenhouse gases, 199

H

High-performance liquid chromatography (HPLC), 63, 64f, 93, 131, 133f, 143, 157, 188f
High-temperature water (HTW), 115, 129
  defined, 1
  dielectric constant of, 5, 5t, 6t
  environmentally benign solvent, 154, 186
  under subcritical/supercritical conditions, 2
HOMO, 103
Hot compressed water, 1
HTW, *see* High-temperature water (HTW)
Hydrogenation of CO_2, 186, 186f
  waste metal as reducing agent for, 135, 135t
  by water splitting with Al, 132–135
Hydrogen bonding, 6–11, 7f–11f; *see also* Water under high temperature and pressure (WHTP) conditions
Hydrogen–deuterium exchange, 8

Hydrogen production; *see also* Hydrothermal water splitting
  with Al, 38–39
  with boron, 43
  with iron, 52–55, 54t
  with iron, assisted by carbonate ions, 55–57
  with metals, 48–52, 49t, 51t, 52t
  with Mg, 43
  by redox of iron oxide with metal additives, 57, 58t, 59t
  by water splitting with Al, 131–132, 132t
  with Zn, 40–42
Hydrogen transfer, 29
Hydrogen transfer reduction of $NaHCO_3$
  with glycerine, *see* Alcohol-mediated reduction of CO_2/$NaHCO_3$ into formate
  with isopropanol in HTW
    hydrogen-transfer reduction mechanism, reaction pathway on, 167–168
    NaOH concentration, effect of, 165–166
    quantity of $NaHCO_3$, 164–165, 165t
    reaction products from isopropanol/$NaHCO_3$, 163
    reaction temperature, effect of, 163–164
    reaction time, effect of, 166–167
Hydrolysis, 11–12, 11f; *see also* Water under high temperature and pressure (WHTP) conditions
Hydrothermal (terminology), 2
Hydrothermal and electrochemical reduction, combination of, 202
Hydrothermal chemistry, 129
Hydrothermal conversion of CO_2, 143
Hydrothermal CO_2 reduction with iron
  experimental section
    analytical results of liquid/solid samples, 64–65
    experimental procedure, 63
    materials, 63
    product analysis, 63–64
  Fe oxidation
    proposed pathway for, 68–69
    reaction conditions on, 65–68, 66t
    formic acid yield, reaction conditions on, 69–72
  overview, 61–63
  proposed mechanism, 72–75, 73t
Hydrothermal reactions and photocatalytic reduction, combination, 201–202
Hydrothermal water splitting
  global hydrogen consumption per industry, 47–48, 48f
  hydrogen production
    with Al, 38–39
    with boron, 43

  with iron, 52–55, 54t
  with iron, assisted by carbonate ions, 55–57
  with metals, 48–52, 49t, 51t, 52t
  with Mg, 43
  by redox of iron oxide with metal additives, 57
  with Zn, 40–42
  overview, 37
Hydrous hydrazine, 186
Hydroxyacetone, 162
5-hydroxymethyl-2-furaldehyde (HMF), 26
Hydroxymethylfurfural (HMF), 13

I

Infrared (IR) absorption, 100
In-situ characterization of hydrothermal reaction, 25–26; *see also* Catalytic hydrothermal reactions
In situ hydrogen, role of, 72–73, 73t
In-situ Raman spectra, 25
Interfacial chemistry between solid/solution, 23–24; *see also* Catalytic hydrothermal reactions
Intergovernmental Panel Climate Change's report (IPCC), 109
Intergovernmental Panel on Climate Change (IPCC), 47
Intrinsic reaction coordinate (IRC) calculations, 103
Ion product, 2; *see also* Water under high temperature and pressure (WHTP) conditions
Iron, hydrogen production with
  assisted by carbonate ions, 55–57
  technique for, 52–55, 54t
Iron, hydrothermal CO_2 reduction with, *see* Hydrothermal CO_2 reduction with iron
Iron nanoparticles, 86
Isomerization, 12; *see also* Water under high temperature and pressure (WHTP) conditions
Isopropanol, 155, 163

K

Ketone (fructose), 14
Kinetic curve
  of hydrothermal production of formic acid, 83f
  of hydrothermal production of methane, 86f
  for hydrothermal production of phenol, 80f
Kinetic model, 24–25; *see also* Catalytic hydrothermal reactions

Index

L

Lactate formation from glycerine, 171
Lactic acid
  from glucose, 15, 27
Langmuir–Hinshelwood mechanism, 24
Levulinic acid, 26
Liquefaction, hydrothermal, 2
Liquid chromatography-mass spectroscopy (LC-MS) analysis, 156; *see also* Glycerine
Lobry de Bruyn–Alberda van Ekenstein transformation, 12
Low-carbon compounds
  hydrothermal reactions
    from CO_2 to methane, 85–88
    from CO_2 to phenol, 79–81
    from CO_2 to simple carboxylic acids, 81–85
Low-energy metal powders, 51
LUMO, 103

M

Magnesium, hydrogen production with, 43
Manganese and formic acid production
  materials and methods
    experimental procedure, 111
    product analysis, 111–112
  overview, 109–110
  results and discussion
    CO_2 in improving hydrogen production from water, 118, 119t
    CO_2 reduction with Mn, potential of, 112–114, 114t
    formic acid formation investigation, 121–124, 122t
    Mn oxidation in water/Mn_xO_y in formic acid formation, 119–121
    reaction of dissociation of water/high yield of formic acid, 115–118
Manganese oxidation in water, 119–121
Mars–van Krevelen mechanism, 24, 27
Meerwein–Ponndorf–Verley (MPV) hydrogen-transfer reduction, 167
Metals, hydrogen production with, 48–52, 49t, 51t, 52t
Methane, hydrothermal reactions from CO_2 to, 85–88
Methanol
  characteristics of, 141
  formation, 149–150
  synthesis of, from CO_2, 144
Methanol production
  conversion of CO_2 into methanol over Cu nanoparticle
    experimental procedure, 149
    methanol formation, 149–150
  conversion of CO_2 to methanol over commercial Cu powder
    general information, 143
    product analysis, 143
    synthesis of methanol from CO_2, 144
  methanol yield, reaction parameters on, 144–147
  over commercial Cu powder, 147–149, 148t
  overview, 141–143
Michiels's scheme for hydrogen production, 55, 55f
Molecular reaction, 2
Molybdenum sulfide, 27
Mulliken charge of H, 136

N

NaOD, 155
NaOH concentration, effect of, 165–166
Near-critical water, 1
Nickel catalyst, 191, 192f
Nitrogen, compounds containing
  CO_2 reduction with $N_2H_4 \cdot H_2O$, 194–196
  experimental procedure, 187
  optimal reaction parameters of experiment with Ni and ZnO catalysts, 192–194
  overview, 185–186, 186f
  possibility of CO_2 reduction into chemicals with $N_2H_4 \cdot H_2O$, 187–189
  product analysis, 187
  reaction parameters on yield of formate
    with Ni and ZnO catalysts, 192
    with Ni catalyst, 189–192
NMR analysis, 156

O

Oppenauer oxidation, 168, 175
Organic acid, conversion of CO_2 into, 29–31, 30t; *see also* Catalytic hydrothermal reactions
Outer Helmholtz plane, 24

P

Phenol, hydrothermal reactions from CO_2 to, 79–81
pH on formic acid yield, 116f
Phonon calculations, 26
Photocatalytic conversion of CO_2, 128
Polycyclic aromatic hydrocarbons (PAHs), 5
Pyruvaldehyde, 27, 162, 176t

Q

Quantum chemical calculations
  aluminum and formic acid production, 131
  of reaction mechanism, 101–105, 103t
Quantum chemistry calculation, 26; see also Catalytic hydrothermal reactions

R

Raman spectra, 25–26
Raman spectroscopy, 64, 65f
Raney nickel, 154
  catalyst, 168, 175
Reaction heat, 123
Reaction mechanism
  proposition of, 174–175
  testing of, 175–176, 176t
Reaction products from isopropanol/$NaHCO_3$, 163
Reaction temperature, effect of, 163–164
Reaction temperature/time, effects of, 160
Reaction time
  defined, 111
  effect of, 166–167
Reactor materials
  about, 173; see also Abiogenic formate production from CO_2
  catalysis effect of, 160–161
Recyclable carrier of renewable energy, 38
Redox catalysis, 26–27; see also Catalytic hydrothermal reactions
Redox of iron oxide with metal additives, 57, 58t, 59t; see also Hydrothermal water splitting
Redox reactions, 24–25
Rehydration of HMF, 13, 14
Renewable energy, recyclable carrier of, 38
Retro aldol reaction, 14–15, 14f; see also Water under high temperature and pressure (WHTP) conditions

S

Scanning electron microscopy (SEM), 64, 67f, 84, 87f, 94
Self-ionization constant, 2
Semiconductor catalyst, 28
Simple carboxylic acids, hydrothermal reactions from CO_2 to, 81–85
Simulation for hydrothermal vent system, 200–201
Single-metal additives, 57
Solar-driven redox reactions, 91
Solar energy, 201
  conversion, 93
Solar-to-formic acid energy conversion efficiency, 105
Solar-to-fuel conversion, 128, 201
Solar zinc powder, 40
Steam–iron process, 52
Stoichiometric reactants, 42
Sulfur, hydrothermal reactions of, 28–29
Supercritical water, 5, 6t
Syngas, coal-based, 142

T

Teflon reactor, 112, 114t
Thermal conductivity detector, 143
Thermochemical water decomposition, 48
Thermogravimetric process, 40
Total organic carbon (TOC), 63, 93, 111
Transesterification, 141
Transition state (TS), 103, 104f

U

Uncatalyzed transfer hydrogenation of CO_2 with glycerine, see Abiogenic formate production from CO_2

W

Waste metal as reducing agent for hydrogenation of CO_2, 135, 135t
Water catalysis, 10f
Water-catalyzed mechanisms, 162, 175
Water density, 2–3; see also Water under high temperature and pressure (WHTP) conditions
Water filling
  defined, 63, 111
  effects of, 190, 191f
  increase of, 68
  $NaHCO_3$ and, on formic acid yield, 69, 70f
  parameter for CO_2 reduction, 117
Water–gas shift reaction, 7
Water splitting, 40
Water splitting with Al; see also Aluminum and formic acid production
  CO_2 reduction by, 136–137
  hydrogenation of CO_2 by, 132–135
  hydrogen production by, 131–132, 132t
Water under high temperature and pressure (WHTP) conditions
  decarboxylation/decarbonylation, 15–16
  dehydration, 13–14, 13f
  dielectric constant
    about, 3–5, 4f
    HTW, 5–6, 5t, 6t
  hydrogen bonding, 6–11, 7f–11f
  hydrolysis, 11–12, 11f

ion product, 2
isomerization, 12
overview, 1–2
retro aldol reaction, 14–15, 14f
water density, 2–3

X

X-ray diffraction (XRD), 64, 65f, 85, 85f, 133f
X-ray photoelectron spectroscopy, 73, 74f
X-ray powder diffraction (XRD), 53, 145f, 189

Y

Yield of products, defined, 167

Z

Zeolites, 80
Zero point energy correction, 39, 39f
Zinc, hydrogen production with, 40–42
Zinc and formic acid production
    experimental procedure, 93
    formate production, proposed mechanism for, 99–101, 99t
    formate production with $NaHCO_3$ and Zn
        reaction characteristics/optimization of reaction parameters, 94–96
        reduction of $NaHCO_3$, 94
    formic acid production with gaseous CO_2, 97–98, 98t
    materials, 93
    overview, 91–93, 92f
    product analysis, 93–94
    quantum chemical calculations of reaction mechanism, 101–105, 103t
    water splitting for hydrogen production with Zn, 96–97, 97t
    Zn–ZnO cycle and energy conversion efficiency assessment, 105–106
Zinc hydride, 41, 100, 130
Zinc hydrolysis, 42

For Product Safety Concerns and Information please contact our EU representative GPSR@taylorandfrancis.com Taylor & Francis Verlag GmbH, Kaufingerstraße 24, 80331 München, Germany

Printed and bound by CPI Group (UK) Ltd, Croydon, CR0 4YY
08/06/2025
01896985-0005